Lunda Under Belgian Rule

Lunda Under Belgian Rule
The Politics of Ethnicity

Edouard Bustin

Harvard University Press, Cambridge, Massachusetts
and London, England
1975

Library of Congress Cataloging in Publication Data

Bustin, Édouard.
  Lunda under Belgian rule.

  Bibliography: p.
  Includes index.
  1. Shaba, Zaïre — Politics and government.
  2. Lunda (Bantu tribe) I. Title.
DT665.K3B86   967.5'18'02   75-9995
ISBN 0-674-53953-2

Preface

The research on which this study is based was carried on during three separate trips to the former Katanga province (now the Shaba Region) of the Republic of Zaïre on 1967, 1968, and 1972, and during a summer visit to Belgium in 1970, but my curiosity toward the former Lunda empire probably dates back to the period immediately preceeding Congolese independence, when a considerable amount of rather shallow speculation was being generated among expatriate circles concerning the potential affinities between Lunda particularism and Katanga separatism, the predictability of the Lunda and Cokwe ethnic votes, and the interaction between such supposedly known quantities and the political prospects of an aspiring politician of no mean local fame — Moïse Tshombe — at which point mention would inevitably be made of the fact that Tshombe had married the daughter of the Lunda paramount chief, or *Mwaant Yaav*, and could accordingly count on his unqualified backing.

African contacts soon taught me that the truth of the Lunda situation was in fact somewhat different, and that the principal, if not the only reason for the ostensible solidarity between the Lunda "emperor" (as he then styled himself) and his ambitious son-in-law was that the Mwaant Yaav needed Tshombe at least as much as Tshombe needed the Mwaant Yaav. I was made aware of the enduring feud between the family of the paramount chief and that of the rising politician, but it was not until several years later that I learned that from the moment of Tshombe's emergence as the leader of separatist Katanga, the ageing emperor had lived in almost constant fear of being deposed or poisoned at his rival's behest. Meanwhile, even the most thorough studies of the Katanga political scene limited their scrutiny of the Lunda factor to an unquestioning restatement of the alliance relationship between the two men.

At an early stage of my research, I was fortunate enough to be granted a generous access to the provincial archives of Katanga at Lubumbashi. This good fortune was somewhat mitigated, however, by the poor condition and by the state of almost complete disarray of this considerable mass of untapped information, spanning a period of some fifty years. Thus, much of my time at that point was spent piecing together scattered, incomplete, or dilapidated files. Such data as I could assemble were then patiently re-assembled in Boston, and completed by a resort to the archives of the former Ministry of Colonies in Brussels,[1] to local archives preserved in the Lunda region itself, and by a small number of interviews with Lunda dignitaries, active or retired civil servants and missionaries. The largest bulk of information was derived from archival sources, primarily from the Katanga provincial (now Shaba regional) archives. Except where otherwise indicated, therefore, all official documents and administrative correspondence referred to in the course of this study were consulted in the Lubumbashi depository.

As the title of this book and the nature of my sources will suggest, this work was not intended as a study of an African society, or of its traditional political mechanisms, but rather as an inquiry into the workings and effects of alien rule upon an African state. In the process of looking into these questions, however, I had to develop some understanding of the nature of political relations in the specific context of precolonial Lunda, but I claim no special insights into the nature of African power, nor do I boast of having evolved any constructs other than analytical to account for the interaction between rulers old and new.

In many ways, the central character of this study is the Belgian colonial civil service, its attitudes, its procedures, its perceptions and distortions, its preoccupations and stereotypes — in a word, its "policies." But what about its *policy*? Notwithstanding substantial evidence that no imperial power in history ever evolved anything resembling a single, concerted, and consistent colonial policy (let alone applied it), studies of British, or French, or even Portuguese colonial policy have been and continue to be written. What such studies usually offer, however, are discussions of successive or competing doctrines of colonial administration — indirect rule, association, etc. Yet, while such discussions, or loftier speculations about the ultimate goals of colonization, did mobilize a respectable amount of intellectual activity, administrators working in the field were only moderately preoccupied with these considerations, even though they might have definite ideas as to how colonies should be run. As observed by a recent student of French policy: "While the metropole proclaimed vague goals about the French mission of civilization . . . the administrators were permitted to govern according to their whims and inclinations."[2]

While they clearly shared the standard views regarding Europe's "civilizing mission" in Africa, Belgian policy-making circles never fully embraced any of the prevailing doctrines of colonial rule. In this they differed from other colonial powers, but mostly on paper, for neither the British nor especially the French ever fully practiced what they professed to believe in matters of native administration. At the same time, however, a high degree of bureaucratic centralization and a vision of administrative "rationality" borrowed from French models made the Belgian administration far less pragmatic than it liked to think; yet Belgian policies were not rationalized in terms of an overriding conceptual matrix but rather on the basis of Alexander Pope's dictum that "whatever is best administer'd is best." Needless to say, a "good" administration under these terms was simply one which combined minimum cost with maximum effectiveness, but then other colonial powers were operating very much on the same assumptions, though not quite as explicitly.

During the Free State era (1885-1908), administrative praxis was heavily on the side of direct rule. This was undoubtedly due in large part to Leopold's determination to secure short-term returns on the considerable sums he had personally sunk into the Congo venture, by means of coercive and monopolistic practices, but also to the absence of large-scale, hierarchized traditional polities in the central and western portions of the Congo, where Leopold's exploitative policies were applied. Areas where such traditional institutions had been relatively well preserved, such as Lunda or the Zande sultanates, were not brought under effective control until much later and, in the case of Lunda, it was not until World War I that a pattern of "native administration" was actually introduced, by which time the colonial bureaucracy had already been in existence for thirty years.

Lunda was in any case a somewhat atypical area in the Congo, representing as it did the most extensive (if not the best organized) of the Bantu states under Belgian control.[3] Like the Zande sultanates of the northeast, however, it was geographically peripheral to the Congo, with historical affinities to other populations living in neighboring territories. But to an even greater extent than the Zande region, which later became a major center of cotton production, Lunda was economically marginal, particularly in Katanga where it was almost totally overshadowed by the development of mining and of urbanization in what had been, in precolonial days, a comparatively barren and inhospitable area. Thus it was deprived, both in the eyes of the colonial establishment and in relation to the rest of the Congo, of two of the most important elements that contributed to the exceptional status enjoyed by Buganda and — to a lesser extent — by Ashanti in the much smaller colonial territories of Uganda and the Gold Coast.

One of the most vexing problems in dealing with Lunda is the use of the term "Lunda" itself. It came into circulation among Europeans at the time of the empire's maximum extension — or, more accurately, at the time when the fame of Lunda's ascent reached the Portuguese establishments on the coast — and its precise scope was not ascertained until the nineteenth century, at which time the influence and power of the empire had shrunk seriously. Cokwe intrusions during the second half of the century completed the dislocation of what had always been, in any case, a loosely knit structure, and by the time the Congo Free State got around to establishing its first modest post in the area (nearly twenty years after the Berlin conference), the Belgian commander of the expedition could write, upon visiting the ruins of a recently abandoned royal capital: "This is all that remains of the famed Lunda empire." A naive statement, of course, and based in part on the Belgian officer's ignorance of the Lunda traditions of a semi-itinerant court, but it was in other respects not far from the truth: the Lunda state had indeed been for over two decades in a condition of unprecedented turmoil. Yet, serious geographers of the late nineteenth century, such as Elisée Reclus or A.J. Wauters, continued to assign to it an importance more in keeping with its past power than with current reality. Belgo-Portuguese negotiations in 1890-1891 had been held for the purpose of partitioning "Lunda," although the then Mwaant Yaav was at that point in control of no more than one percent of the territory being argued over at the conference table. The fiction was maintained even after the partition, as the Portuguese gave the name of "Lunda" to the portion of the disputed area that was assigned to them in 1891 (i.e., west of the Kasai River), even though its population was predominantly Cokwe.

A different sort of ambiguity exists regarding traditional African usage. From a relatively small nuclear area in the Katanga (Shaba) region of Zaïre, and through a process about which there is little solid information, relatively small bands of migrants exported a form of political centralization to the segmented populations scattered over a wide area to the west, south, and southeast of their own homeland. The result was the creation of a far-flung network of tributary relationships but also, at the periphery, of fully autonomous polities such as Kasanje, the Yaka kingdom, or the Kazembe states which successfully broke off from the parent state without entirely forgetting their origins. But while the term "Lunda" may be used, and has indeed been used, to cover all the populations which were at one time linked through this tributary network, the ethnic distinctiveness of these populations was always perceived by the nuclear Lunda or "Aruund."[4]

Strictly speaking, therefore, it would probably be best to regard the word Lunda as a political-historical term designating a now defunct system, and to distinguish it from "Ruund" in somewhat the same fashion as "Ottoman" is distinguished from "Turk." This was the view suggested nearly forty years ago by some Belgian officials such as Prosper Montenez, who noted in 1936: "The Aluunda, once very powerful, today form a homogeneous group only in the cradle of their race, the area between the Kalagne and Lulua Rivers, at Kapanga. This is Uluunda—Luunda country."[5] In fact, except in specific cases or to avoid ambiguity, the name "Aruund" has not received wide acceptance in administrative, or even in anthropological parlance. Belgian, British, and Portuguese officials continued using the term Lunda rather loosely, and anthropologists such as Merran McCulloch, for example, have found it easier to refer to the northern, eastern and southern Lunda, while recognizing that they are distinct, "both linguistically and culturally."[6]

But there are more practical reasons for us to defer to the prevailing usage in this study. Although the Mwaant Yaav exercised authority on a mere fragment of the former empire by the time colonial administrators established a measure of effective occupation over the area, they restored his jurisdiction (at least nominally) over several regions which had slipped away from his control over the previous thirty years or so. In the process, "Lunda" chiefs—chiefs subordinated to the Mwaant Yaav, whether or not they were actually Aruund—were set up throughout the western portion of Katanga and their nominal allegiance to the Lunda paramount was officially acknowledged. This early attempt to streamline the patterns of native administration in this part of the Congo was reappraised in the early 1920s, when it became obvious that those tribes which had shattered the old empire (the Cokwe), or had recovered their autonomy as a result of its collapse (e.g., the Lwena) accepted only grudgingly the authority of these "Lunda" chiefs. A handful of Cokwe and Lwena chiefs were granted recognition in turn, but only when this proved to be absolutely inevitable. As a result, a majority of the chiefs of western Katanga continued to owe at least nominal allegiance to the Mwaant Yaav, and the area corresponding roughly to the three colonial *territoires* of Kapanga, Sandoa and Dilolo (plus a portion of Kolwezi) was commonly referred to as "Lunda country" throughout the colonial period, even though the Aruund proper were a distinct minority in the region as a whole.

Again, it is useful to bear in mind that "Lunda" describes the collectivity of those people living under the authority of chiefs who acknowledged (or were pressured by the colonial administration into acknowledging) the Mwaant Yaav's paramountcy. Over the years, however, the specific obligations deriving from this sort of allegiance became increasingly diluted and, since they were regarded as binding only under customary law, their performance was not enforced by the Belgian authorities. The sense of "Lundahood" did not disappear, however, and villagers were commonly classified as "Lunda" by government officials, or even by anthropologists, simply by virtue of the fact that they lived under the jurisdiction of a "Lunda" chief.[7]

During the late 1950s, against the context of a partly improvised decolonization, the notion of a "Lunda Empire" was revived by a number of personalities associated, for the most part, with Moïse Tshombe, and the concept was also taken up, with a slightly different instrumentality, by the then Mwaant Yaav, Ditende Yaav Naweej III. Although the precise political implications of this projected revival were never clearly spelled out, its territorial scope was understood to coincide with the lands controlled by Lunda chiefs and, in some bolder versions, with all of western Katanga.

In this study, I have used the term "Lunda heartland" to describe the area extending west of the Lubilash River to the Kasai River, and on a north-south axis, from the lands of the Asalampasu, BaKete, and Bena Kaniok (i.e., the former province of Kasai) to the Angolan and Zambian borders.[8] This area corresponds approximately to the three above-mentioned colonial subdivisions of Kapanga, Sandoa, and Dilolo. Of these three units, only the first was placed under the Mwaant Yaav's direct authority during the colonial period. Within its boundaries are located the traditional sites of initiation and burial of the successive rulers of Lunda, in the Nkalaany River Valley. It represents the original nucleus of Lunda dispersion and the empire's last stronghold against the encroaching Cokwe. It also roughly coincides with the area which the Aruund had managed to reconquer from the Cokwe by the first decade of this century, at which time effective Belgian occupation put an end to open warfare — or, to be more accurate, when "pacification campaigns" were substituted for so-called "tribal wars."

No such ambiguities exist regarding the use of the term "Cokwe," for which I followed the spelling preferred by modern Africanists, rather than any one of the dozen or so versions of the name used by British, Belgian, or Portuguese writers (Chokwe, Tshokwe, Tshiokwe, Quioco, Kioko, Tshok, Badjok, Tutshiokwe, Achoku, etc.), which will appear only in direct quotations. Such variations in spelling also occur in reference to the title of Mwaant Yaav (Mwata Yamvo, Mwata Yamfu, Mwant Yavu, Muatianvo, Mwata Yambo, etc.), as well as in the titles of the senior dignitaries of the Lunda court — the Nswaan Mulopw, Rukonkesh, Nswaan Muruund, Kanampumb, etc. — and the same guidelines have accordingly been followed. When dealing with other chiefs, however, I have adhered to Belgian administrative usage, primarily because their titles were also officially used to designate the native administration units placed under their jurisdiction.

One last remark concerning nomenclature. Since 1965, and particularly in 1971 and 1972, a vast number of name changes have occurred in the former Belgian Congo. The Congo is now known as Zaïre, Katanga as Shaba (Swahili for "copper"), and Elisabethville as Lubumbashi, while Léopoldville, Stanleyville, Jadotville and other cities have reverted to their earlier names of Kinshasa, Kinsangani, Likasi, etc. Former administrative divisions such as provinces or districts have been renamed *Régions* and *Sous-régions*, while individual citizens have been enjoined to use African names exclusively. To avoid unnecessary confusion, however, names and titles used in this study are those that were current during the time period covered.

Acknowledgments

In the process of writing this book, I have incurred a debt of gratitude toward many individuals in Zaïre, in Europe, and in the United States and I should like to express my thanks, however inadequately, for the various forms of assistance which I received at every stage during the preparation of this study.

I am indebted to the Zaïrois staff of the former Direction du Service Territorial of Katanga (Shaba), at Lubumbashi, for providing me with access to the provincial archives, and to M. Grandjean, of the Belgian technical assistance mission, for guiding my first steps into this intricate mass of documentation. The late Mwaant Yaav, Muteb II Mushid, was extremely helpful in his dual capacity as traditional chief and field official by giving me access to local archives at Kapanga and by providing me with an introduction to several court dignitaries. I am particularly indebted to the American and Zaïrois members of the Methodist mission at Musumba, not only for their hospitality and for the logistical support they gave me, but also for illuminating insights they contributed to my research. Special thanks, in each of these respects, should go to the Reverend William Davis, and to his son whose command of the Ruund language was of invaluable assistance in conducting interviews.

In Belgium, Mesdames Madeleine Van Grieken-Taverniers and Olga Boone helped me uncover new materials in the archives of the former Ministère des Colonies and in the collections of the Musée Royal de l'Afrique Centrale, respectively. Two young scholars whose own research on Lunda is still in progress, Ndua Solol (Edouard), of the Université Nationale du Zaïre, and J. Jeffrey Hoover, of Yale University, read early versions of my manuscript and contributed many valuable comments, while letting me share some of their own findings.

At various points, Jan Vansina, of Leuven University, Roger Anstey, of the University of Kent, and Jean-Luc Vellut, of the Université Nationale du Zaïre, read extensive portions of my study and helped sharpen my interpretation of the historical material I was trying to handle. Robert I. Rotberg, of MIT, and M. Crawford Young, of the University of Wisconsin, contributed their highly constructive comments to a subsequent version of the manuscript. Although I trust that every one of them will be able to find some trace of their valuable suggestions in my finished product, it goes without saying that they are by no means responsible for the selection and interpretation of my data, the burden of which is entirely my own.

Andrea Truax typed and retyped my manuscript with a unique combination of patience and diligence for which a grateful acknowledgement hardly seems appropriate.

I benefited from several forms of institutional support from Boston University, particularly from a research grant of the Graduate School, and from different forms of concrete encouragement by the African Studies Center. I cannot, however, divorce such corporate support from the warmer and far more personal form of interaction which it concretely entailed on the part of my colleagues at the Center, particularly the late Alphonso A. Castagno, Creighton Gabel, and Norman R. Bennett, who all helped in many different ways.

Finally, a special expression of appreciation mixed with apology must go to Francine, Denis, and Olivier for the many hours we could not share, but also for all those we did share because of this book.

Contents

Tables

Maps

Lunda Under Belgian Rule

Abbreviations

AMC   Archives du Ministère des Colonies. AMC documents are further identified by reference to the collections inventoried by the officials of the former minister of colonies, as indexed in E. Van Grieken and M. Van Grieken-Taverniers, *Les archives inventoriées au Ministère des Colonies* (Brussels, ARSC, 1958). (The archives of the former ministry of colonies have now been formally absorbed by the archives of the Belgian Foreign Office, but their functional and physical distinctiveness has been maintained).

AIMO   Affaires Indigènes et Main d'Oeuvre.

APK   Archives Provinciales du Katanga.

ARSOM   Académie Royale des Sciences d'Outre-Mer, formerly known as Académie Royale des Sciences Coloniales (ARSC) and Institut Royal Colonial Belge (IRCB).

ATD   Archives du Territoire de Dilolo.

ATK   Archives du Territoire de Kapanga.

ATS   Archives du Territoire de Sandoa.

BCK   Chemin de Fer du Bas-Congo au Katanga.

BTK   Bourse du Travail du Katanga, later replaced by Office Central du Travail au Katanga (OCTK).

CEPSI   Centre d'Etudes des Problèmes Sociaux Indigènes (Lubumbashi), later renamed Centre d'Etudes des Problèmes Sociaux Congolais.

CI   Circonscription Indigène.

CACI   Caisse Administrative de Circonscription Indigène.

CFL   Chemin de Fer des Grands Lacs.

CSK   Comité Spécial du Katanga.

CRISP   Centre de Recherche et d'Information Socio-Politiques (Brussels).

DST   Direction du Service Territorial.

KDL   Katanga-Dilolo-Léopoldville (Railroad).

MRAC   Musée Royal de l'Afrique Centrale, Tervuren (Belgium).

RUFAST   Recueil à l'Usage des Fonctionnaires de l'Administration en Service Territorial.

UMHK   Union Minière du Haut-Katanga.

1

Cultural Diffusion and Political Expansion: The Growth and Decline of the Lunda State

Lunda occupies a unique position in the political history of Central Africa. Culturally, it was in all probability less important than the Luba crucible to the northeast and certainly less innovative. Nor, until further research clarifies our views on the subject,[1] can we ascertain the precise importance of Lunda in the economics of long-distance trade, other than to assume that it must have fuelled to some significant extent the commercial activities of Kasanje, and later those of Kazembe. In the absence of any substantial evidence, we may hypothesize that commerce acted as a catalyst in the emergence of state organization in South Central Africa, as it did in other parts of the continent, but such a process would not in itself be particularly original.

Lunda's particular contribution to the political history of South Central Africa was the diffusion of its governmental patterns over an area ranging from Angola to Malawi. So crucial was this phenomenon that Jan Vansina could argue convincingly that

> the crucial event in the earlier history of Central Africa has been not the creation of a Luba kingdom . . . but the introduction of Luba principles of government into Lunda . . . and their transformation by the Lunda. The new political pattern which evolved around 1600 in the Lunda capital could be taken over by any culture. Its diffusion was to condition until 1850 the history and the general cultural evolution of a huge area. Even now its effects on the peoples of Central Africa are still discernible.[2]

This being said, the remarkable spread of the Lunda political network cannot be easily accounted for. A combination of factors, some having to do with Lunda concepts and attitudes regarding land and political relationships, and some pertaining to the physical character of the area in which their expansion took place, may be viewed as having played an essential role.

In the first category, the central concepts were, on the one hand, the clear-cut distinction between rights in land and authority over men (or, as one author puts it, between *dominium* and *imperium*[3] and, on the other hand, the interacting notions of positional succession and perpetual kinship.

The basic unit of the Lunda system was the *ngaand*, usually translated as "village" but representing a human community as well as the land area it occupied. It was ruled—or perhaps, more accurately, represented or even symbolized—by the *mwaantaangaand* whose name was identified with that of the *ngaand* itself and whose position technically depended on matrilineal descent from the original settler of the land. His duties were primarily ritual and narrowly related to the use of the land (fertility rites, timing of communal tasks, etc.) or to the propitiation of deceased ancestors. His failure to perform effectively in this role as conciliator of unseen forces—as evidenced, for example, by repeated crop failure or continuing drought—could lead to his abdication, and his authority over mundane matters was in any case circumscribed by that of a council of elders or *ciyul*.[4]

Whether peaceful or forcible (and we shall see that there is considerable ambiguity on this point), the integration of the various villages into the Lunda political network took the form of the dispatching to a given area of a political administrator, the *cilool* (*kilolo*), whose primary role was that of a tax collector and who had no authority over the land itself—at least *qualitate qua*. A *cilool* could and often did control land in the area under his jurisdiction but only when such property had been transferred (whether voluntarily or not) by one or several *mwaantaangaand*,[5] and this factor did not affect his role as an intermediary between the villages and the royal court. If the tribute was not collected or forwarded regularly, an inspector or *yikeezy* (*ikeji*) could be sent from the court to reside in the *cilool's* village and control—but not replace—him. *Tukwata* (sing., *kakwata*), who combined the functions of tax collectors and court messengers, were also used to expedite the forwarding of tribute.

The two concepts of *dominium* and *imperium* converged at the highest level in the person of the *Mwaant Yaav* or king, who combined in the highest degree the ritual functions associated with land (in his capacity as *Mwiin mangaand,* i.e., owner of all the *ngaand*) with those political functions represented at the local level by the *cilool* and, even more explicitly, by the *yikeezy.* This dichotomy was represented by the structure of the royal court, which consisted of three groups of dignitaries.

A first group included titleholders whose position was linked to the principle of kingship itself: the *Nswaan Muruund* (*Swana Mulunda*) or perpetual mother of the Lunda; the *Rukonkesh* (*Lukonkesha*), positional mother of the king; the *Nswaan Mulopw* (*Swana Mulopwe*), positional son of the king—and thus often described as "crown prince," although his succession to the throne was by no means automatic, etc. These dignitaries were entitled to the direct income of certain tributary areas and, as such, controlled their own *cilool* who were regarded as their positional sons. Of the other two groups of dignitaries, it might be said, in a somewhat oversimplified way, that one symbolized all the religious concepts associated with the land while the other was representative of the political, tribute-oriented structure. The first of these two groups included the fifteen *Acubuung* (*Tubungu*) or *mwaantaangaand* of the fifteen original villages of Lunda in the valley of the Nkalaany (Kalagne) or upper Bushimaie River, as well as such ritual figures as the *Mwanamutombo,* the king's physician; the *Mwadi,* keeper of the royal tombs; or the *Muinda,* guardian of the throne, etc. The ritual and land-related base of the Mwaant Yaav's power was evidenced by the central role played by the fifteen *Tubungu* in the selection, initiation, and investiture of the king.[6] The other group consisted of tributary chiefs who were responsible for the administration of certain specific areas (e.g., the border areas or "marches" or Lunda) whether by virtue of a direct appointment, or as chiefs of semiautonomous "buffer" areas who had sought or accepted the Mwaant Yaav's protection. These latter dignitaries did not reside in the royal capital (*Musuumb*) but were represented at court by the *Ntomb* (whose original position may possibly have been that of hostages). The *Ntomb* transmitted the tribute from the chiefs and other tax-collecting officials that they represented.[7]

The stability and continuity of the whole system was further reinforced by the generalized use of perpetual kinship, which fictionally maintained the original blood ties between various titleholders, and of positional succession, which amounted to identifying any person holding a position of authority with the original holder of the title.[8]

The positive advantages of these various concepts in terms of the rapid expansion of the Lunda political network are obvious. The dichotomy between "land chiefs" and "political chiefs" made it possible to assimilate local rulers into the tributary network as *mwaantaangaand* while *cilool* functions were performed by the Lunda intruders. At the same time, the principles of positional succession and perpetual kinship could be superimposed over the local descent systems without disrupting them. As a result, the conflictual potential of local traditions and values was minimized, and assimilation into the Lunda political network was made relatively easy to accept.

In many ways, the remarkable expansion of the Lunda political system during the seventeenth century is reminiscent of the spread of a creeping vine, rapidly shooting its tendrils in several directions and wrapping them around obstacles rather than attempting to topple them outright. Yet, it would be equally misleading to assume that Lunda expansion did not involve any significant amount of organized violence. The primary purpose of direct military operations, however, seems to have been the procurement of slaves to be used as laborers on domestic plantations or, as commercial contacts with the Portuguese were being developed, as exportable commodities. The same ambiguity surrounds the concept of tribute. The operation of the tributary network exhibits all the functional characteristics of an economic process. The commodities transmitted to the royal capital as tribute (salt, copper, slaves, and later ivory) were eminently exportable over the long-distance trade routes.[9] In return, the *Musuumb* sent to the tributary chiefs "gifts" of imported cloth, beads, luxury goods, and later an occasional firearm. Yet, although tribute was clearly essential in fueling the long-distance trade, its commerical significance was apparently not perceived as central. Vellut notes that "the oligarchy did not regard themselves as traders."[10] The intrinsic or exchange value of the tribute being routed to the capital along the extended lines of political control was probably less important than the symbol of political dependency represented by the act of transmission itself. Thus, while it may be an exaggeration to claim—as disapproving Belgian officials often tended to do—that the Lunda empire existed only for the purpose of collecting tribute, one might nevertheless assert that tribute was its very lifeblood, not so much because of its economic importance but because its very circulation kept the arteries of the empire open.

While the ambiguity of such a system made it eminently adaptable, and capable of expanding rapidly over vast areas, it also accounted for its singular fragility. Even its rather limited functions, to the extent that they were organized around a center (in this case, the royal capital), tended to become diluted in the peripheral regions of the empire. Vansina observes that "the outer provinces could do as they pleased so long as tribute was being paid."[11] But the tribute was not always being paid—which is another way of saying that the royal court's monopoly of long-distance trade was often circumvented; it was in this fashion that the sclerosis of the system first manifested itself openly. If the chief of an outlying area "forgot" to send the tribute for a number of years, the Mwaant Yaav could dispatch a *kakwata* and, failing that, a military expedition could be mounted, but as the spatial range of the tributary network expanded, such expeditions became too cumbersome even to attempt and a de facto lapse of the tributary relationship could ensue. Thus the very scope of the empire's territorial spread worked against the effective operation of its most essential politico-economic function, at least if it tended to exceed a certain optimum dimension.

Over the years, the tributary system proved increasingly inadequate to accommodate the growing volume of commercial transactions and to adjust to the increased availability of imported goods. The crude, unilateral form of "pricing" enforced by the royal court through its practice of rewarding tributaries with "gifts" of imported goods was increasingly perceived as unrealistic, and encouraged local chiefs to initiate direct commercial contacts with the enterprising African traders who were beginning to reach Lunda over new southerly routes around the turn of the nineteenth century. Partly because of its eccentric location with respect to these new trade routes, the Musuumb was powerless to prevent such contacts, and the tribute was voided of its economic significance before it was totally disrupted by the Cokwe invasions during the last quarter of the century.

If, as evidenced by the Kazembe of the Luapula and by many other examples, the peripheral areas of the empire were in constant danger of spinning right out of its orbit, and if military force was not forthcoming to restore tributary relations, then it is legitimate to inquire how such relations were imposed in the first place. Some elements of an answer may be gathered from the oral traditions of the Lunda themselves. These would appear to suggest that much of the initial expansion toward the west and southwest in the seventeenth century was in the form of migrations involving segments of the original population (under circumstances which remain far from clear) and that such migrations, stretching over great distances, took place into rather infertile and thus very sparsely populated areas. At the same time, these same traditions recount the perennial and inconclusive battles sustained by every Mwaant Yaav to secure the boundaries of the Lunda heartland against their immediate neighbors to the north and east—the Nkongo, the Salampasu, the Kete, the Kaniok, the Bena Kalundwe of Mutombo Mukulu, etc. Actual territorial expansion in that direction was slight but the primary purpose of these campaigns against culturally alien groups may have been simply the acquisition of slaves rather than the establishment of direct political control. By contrast, organized military expeditions dispatched toward the southeast for the apparent purpose of controlling salt and copper deposits represent yet a third form of Lunda expansionism, resulting however in the creation of peripheral power centers which rapidly assumed virtual political autonomy, even though they retained the internal principles of Lunda state organization.[12]

A summary overview of Lunda historical traditions will help clarify some of these points. Oral tradition indicates as the cradle of the Lunda nation (indeed, of all mankind) an area in the valley of the Nkalaany, or upper Bushimaie, River no more than twenty miles east of the present site of the royal capital (not surprisingly, since for ritual reasons the successive sites of the Musuumb were always located in the vicinity of the nation's "birthplace"), but also nearly as close to the traditionally hostile lands of Lunda's northern neighbors. The same sources then trace descent through several generations of brother-sister marriages[13] to the man who may be said to serve as a bridge between the myths of origin and the "historical" past, Mwaaku (see Table 1). There is, in fact, considerable uncertainty as to whether the name Mwaaku represents a single individual, a combination of two or more successive persons, or what David P. Henige characterizes as "an archetypal example of epoch personification,"[14] but there is agreement that the said Mwaaku ruled only as a *primus inter pares* among the original *mwaantaangaand* or *tubungu,* of whom the fifteen *tubungu* of the Mwaant Yaav's court (see above) are thought to be the positional successors. Mwaaku's successor and descendant,[15] Nkond (or Konde), had two sons, Cinguud (Kinguri, Tshinguli) and Cinyaam (Cinyama, Tshiniama), as well as a daughter named Rweej (Lueji). As a result of a quarrel with his two sons (the details of which may vary somewhat according to the source but are of no particular importance to us), Nkond reportedly decreed that Rweej should succeed him. After Rweej had begun to reign, one Cibind Yiruung (Cibinda Ilunga, Tshibinda Ilunga), brother of Ilunga wa Lwefu, ruler of the second Luba "empire," wandered into the land of the Lunda at the head of a hunting party and so impressed the queen that she made him a consort ruler. What this myth (with its detailed literary embellishments) almost certainly attempts to euphemize is the extension of Luba power over the area. Tradition has it that the two brothers of Rweej, Cinguud and Cinyaam, who had not objected to her becoming queen, refused to pay homage to the newcomer (not to mention the Luba entourage he seems to have introduced with him) and migrated in different directions.

The story of these migrations itself is a rather tangled one. For one thing, it seems that migrations, as opposed to mere hiving off caused by population increases, had taken place before. For another, instead of two migratory thrusts led by the two brothers, there were probably several, none of them involving large numbers. The best known of these migrations, that of Cinguud, eventually led to contacts with the Portuguese, either in the immediate hinterland of Luanda or at the coast, and to the foundation of the kingdom of Kasanje (after the name of Cinguud's nephew), in which the Lunda element was rather thinly diluted, and whose population was referred to under the name of Imbangala. Although Kasanje soon acted as the major relay for the introduction of American crops and European goods into Lunda and successfully tried to discourage direct Portuguese trade with the Mwaant Yaav, the kingdom never fell within Lunda's orbit and, as such, would be of no direct interest to us except for the fact that it provides our first tentative correlations with Western chronology; these suggest that the migrations (and consequently Cibind Yiruung's ascent to power in Lunda) took place between the mid-sixteenth century and, at the latest, the first decade of the seventeenth century.[16]

For his part, the second brother, Cinyaam, migrated to the south into the area situated between the upper Kasai and upper Zambezi, there to found the Lwena kingdoms. This second thrust is far less well

Table 1. Comp

Dias de Carvalho (1890)	Duysters (1927)	Biebuyck (1955)
	(Tshinaweji) (God)	Mbar Cinawezi = Musaan
	Mwaku	Mwaku = sister
Iala Macu = Condi	Yala Mwaku	Yaal = sister
		Matit = sister
	Konde	Nkond = sister
Lueji = Ilunga	Lueji = Tshibinda Ilunga	Rweej = Cibind Yiruung
Ianvo	Lusenge Naweji	Yaav a Iruung
Noeji Ianvo	Mwanta Yavo Naweji	Nawezi I
Muquelenge Mulanda	(Mukelenge)[a]	
Muteba	Muteba (end 17th.-beg. 18th.c)	
Ianvo Noeji		
Mucanza		
Mulaji		
Umbala		
Ianvo (2d half 18th.c.)	Yavo ya Mbanyi (beg. 19th.c.)	Yaav II
(Quicomba)[a]	Tshikombe Yav (abdic.)	Cikomb a Yaav
Noeji a Ditenda (?-1852)	Naweji ya Ditende	Nawezi a Ditend
Mulaji Umbala (1852-1857)	Mulaji a Mbala (1852-1857)	
Cassenquene (1857)	Tshakasekene Naweji (1857-58)	
Muteba (1857-1873)	Muteba ya Tshikombe(1858-73)	
Umbala (1874)	Mbala (1874)	Mbar
Noeji Ambumba (1874-1883)	Mbumba (1874-1882)	Mbumba a Mulazyi a Mpembe
Chibinda (Ditenda) (1883-84)	Tshimbindu (1883)	Cimbindu
Cangapua Noeji (1884)	Kangapu (1884)	Kangapu Nawezi
Muriba (Quimbamba) (1884-86)	Mudiba (1884-1885)	Mudiba
Mucanza (1886-1887)	Mukaza Mutanda (1995-1887)	Mbar
Umbala (1887-1888)	Mbala (1887)	Mutanda Mukaaz

[a]Did not reign.

of Lunda rulers.

Vansina (1960)	Ngand Yetu (1963)	Nsaang Ja Aruund (1969?)
	Chinawej Mbar	Cjinawej Mbar (ca.1500)
Mwaakw	Mwaku (ca.1500)	Mwaku a Mbar
	Iyal a Mwaku	Iyal a Mwaku
Nkond	Nkond a Matit	Nkond a Matit (ca.1600)
Rweej = Cibind Yiruung	Rweej a Nkond = Cibind Yirung	Ruwej = Chibind Yirung
(1600-1630)	(after 1600)	
Luseeng (1630-1660)	Yavu a Yirung (1625?)	Yavu a Yirung
Yaav Naweej (1660-1690)	Yavu a Nawej	Yavu a Nawej (ca.1650)
Muteba (1690-1720)		
	Mbal Iyavu	Mbal Iyavu
⎧ Mulaji	Mukaz Munyingakubilond	Mukaz Munying Kabalond
⎪ Mbala (1720-1760)	Muteb a Kat a Kanteng (?-1750)	Muteb a Kat Kanteng
⎩ Mukanza	Mukaz Waranankong (1750-1767)	Muteb a Ranakong
Yaav ya Mbany (1760-1810)	Naej Mufa mu Chimbundj (1767-1775)	Nawej Mufa Muchimbundj
Cikombe Yav	Chikomb Iyavu Italesh (1775-1880)	Chikombi Iyavu
Naweej II (1810-1852)	Nawej a Ditend (1800-1852)	Nawej Ditend
Mulaji II a Mbala (1852-57)	Mulaji a Namwan (1852-1857)	Mulaj a Namwan
Cakasekene (1857-?)		
Muteba II ya Cikomb	Muteb a Chikomb (1857-1873)	Muteb a Chikombu (1852-73)
Mbala II (1874)	Mbal a Kamong Isot (1873-1874)	Mbal a Kamong Iswot (1874)
Mbumba (1874-1883)	Mbumb Muteb a Kat (1874-1883)	Mbumb Muteb a Kat (1874-82)
Cimbindu (1883)	Chimbindu a Kasang (1883-1884)	Chimbindu a Kasang (1883)
Kangapu	Kangapu Nawej (1884)	Kangapu Nawej (1884)
Mudiba	Mudib (1884-1886)	Mudib (1884-1885)
Mukanza (?-1887)	Mutand Mukaz (1886-1887)	Mutant Mukaz (1885-1887)
Mbala III	Mbal a Kalong (1887)	Mbal a Kalong (1887)

Colonial and post-colonial period (1887-1975)

Mushid I (Mushidi) (1887-1907) (d. 1909)

Muteb a Kasang (Muteba) (1907-1920) (invest., 1916)

Kaumb (Kaumba) (1920-1951)

Ditend Yavu a Nawej III (Mbako Ditende Yawa Naweji)
   (1951-1963)

Mushid II Lumanga Kawel (Mushidi Gaston) (1963-1965)

Muteb II Mushid (Yav "Tshombe" David) (1965-1973)

Mbumb II Muteb (Tshombe Daniel/Muteb Dipang)
   (1973-   )

documented than that of Cinguud: for obvious reasons there is no possibility of cross references to Western sources, and the oral traditions
themselves are contradictory. Van den Byvang notes that the old
Lunda tales refer only to Cinguud,[17] and the Lwena identify Cinyaam
as a descendant of their founding ancestor who had allegedly left the
Lunda heartland before Cibind Yiruung's arrival.[18]

Other groups which seem to have departed at or about the same
time are said to have been the ancestors of the Cokwe, the Minungu,
the Shinje, and the Songo, although here again there is considerable
uncertainty as to whether or not these migrations are part of the two
major currents of dispersion identified above.[19] What clearly emerges
from the conflicting oral traditions is that, beginning in the first half
of the sixteenth century, a continuous stream of Lunda migrants
moved westward from the tribal heartland until they had gradually
occupied or brought under control the entire area between the middle
courses of the Kasai and Kwango rivers. This movement continued
until the great Cokwe invasions of the late nineteenth century. In the
process, a number of autonomous Lunda dynasties were set up
throughout the area: the three principalities of the Kapenda line
(Kapenda ka Mulemba, Kapenda ka Malundo, and Kapenda
Masongo) east of Kasanje; Mai Munene near the confluence of the
Kasai and Luachimo rivers; Mwata Kumbana on the Loange River;
and, of course, Mwene Mputu Kasongo, first of the line of Lunda
rulers who founded the Yaka kingdom on the Kwango River and
assumed the title of *Kiamfu*. While these several states were or became
independent from the Musuumb, other groups who settled closer to the
Lunda heartland remained within its orbit and organized the western
and southern marches of the Lunda state. Thus originated the two
Kahungula titles (one on the west bank of the Tshikapa, the other on
the Luembe River), the Bungulo, the Sakambundji of the upper
Kasai, and others.[20]

Yet another migration that reportedly took place during the reign of
Cibind Yiruung involved some of his Luba followers who, after considerable peregrination, founded the Bemba nation and settled between lakes Tanganyika, Mweru, and Bangweolu. In that particular
instance, it is almost impossible to identify the origins of the Bemba as
either Luba or Lunda. As with the cases of Kasanje, Mwene Mputu
Kasongo, and others, however, it should be noted that while this
migration represents another instance of cultural diffusion, it did not

imply an extension of political control from the Lunda heartland over the far-flung areas in which these various groups eventually settled.[21]

The construction of the Lunda empire proper took place during the next hundred years under the successors of Cibind Yiruung. According to the tradition prevailing at the Musuumb, Rweej was childless[22] and Cibind Yiruung took a second wife, one Kamonga Lwaza, whose son became king under the name of Yaav a Yiruung or Luseeng.[23] Luseeng is credited by Duysters with having established the bases of Lunda military organization and surrounded his capital, which he named Musuumb (Musumba), with the type of fortification (moat and palisade) that characterized all Lunda settlements until the advent of colonial rule. He also reportedly created the court titles of *Nswaan Mulopw* (*Swana Mulopwe*), *Kalala, Kanampumba, Mwadi Mwishi,* and created for his mother the dignity of *Rukonkesh,* all of which became permanent. He was killed in war against the Kaniok and was succeeded by Yaav Naweej (Yavo Naweji) whose name, combined with the generic title of *Mwaant* became *Mwaant Yaav,*[24] thereafter adopted by all his successors. Yaav Naweej, who appears to have reigned around the middle of the seventeenth century, pursued a deliberate policy of expansion and consolidated court organization. The Lunda practice of not dispossessing local notables, and of confirming their positions as land chiefs while exacting tribute through a *cilool,* facilitated the integration of autonomous groups into the tributary network and the boundaries of Lunda were extended westward to the Kasai.

It appears probable that the success of this policy was due in large part to the fact that the populations thus subjected to tribute were direct descendants of the nuclear Lunda groups who had gradually moved to the west in search of arable land. This view may be indirectly confirmed by the failure of Yaav's attempts to extend his control over the Cokwe and the Lwena. Similarly, expeditions directed against northern neighbors met with rather indifferent success. Some ground was gained at the expense of the Kete (although the king was captured during one of these forays), but other groups proved more resilient and Yaav Naweej actually lost his life during a campaign against a northern tribe.[25]

The end of Yaav Naweej's reign inaugurated a period of instability during which the descendants of Cibind Yiruung seem to have been challenged by a pretender descended from a sister of Rweej,[26] one Mukelenge Mulanda Kasekele (or Kaseka), founder of the Mwene

Mpanda line, who was eventually defeated and killed but whose followers may have migrated eastward to the upper Lualaba. Another
potential successor, Kabeya, also descended from Mwaaku, had already been set aside, according to Duysters, because of his unseemly
conduct during the ill-fated expedition which had cost Yaav Naweej
his life.

There are serious factual and chronological discrepancies among
the various sources regarding the next hundred years or so of Lunda
history. The earlier authors (Dias de Carvalho, Van den Byvang, Duysters), followed by Vansina, identify a number of successive rulers who
are described by Duysters as "brothers" and who are succeeded in turn
by one Yaav ya Mbany.[27] Oral traditions compiled more recently
under the auspices of the Methodist and Catholic missionaries based at
the Lunda royal capital, however, offer significantly different data
(see Table 1), and Biebuyck ignores this period entirely in his otherwise incomplete listing of Lunda rulers.

The names of Muteba (Muteb a Kat Kanteng) and that of Mukanza
(Mukaz Waranankong) appear in the conflicting chronicles, although
the dating of their respective reigns is far from precise. Muteba is
credited by Duysters with a resumption of military operations against
Lunda's northern neighbors and with a renewed expansion of the
empire to the southwest, but his reign is also associated with two
further sets of conquests and/or migrations, which may well have been
motivated by the political instability that apparently prevailed during
the first half of the eighteenth century.

The first thrust was that of the Kazembes which eventually resulted
in the foundation of two semiautonomous kingdoms: one on the upper
Lualaba (Kazembe of the Lualaba), which controlled salt and copper
deposits, and the other on the Luapula (Kazembe of the Luapula), a
much more extensive state which later entered into contact with the
Portuguese establishments at Tete and reached its maximum
expansion between the mid-eighteenth and the mid-nineteenth centuries.[28] The second group of conquests, beginning during Muteba's
reign and continued after his death, took place in a southerly direction and involved three Lunda adventurers, Musokantanda, Kanongesha, and Shinde, each of whom carved out for himself a domain in
the upper basin of the Zambesi. Musokantanda subjugated the unorganized Kaonde. Kanongesha established himself over a predominant-

ly Mbwela population in the area of Mwinilunga (Zambia), and Shinde (or Ishindi) settled farther to the south.

Under Muteba's successors, the feud with the rival line of Kabeya, which had started at the death of Yaav Naweej, continued unabated and, according to Duysters, Muteba's brothers, Mulaji and Mbala, were both killed by Kabeya's followers who seem to have included a large proportion of Kete and/or Kaniok and were based to the east of the Kasidishi River—i.e., within miles of the Nkalaany Valley, the ritual center of Lundahood. Mwaant Yaav Mukanza had to recognize the virtual autonomy of the Kazembe and created for his grandnephew, Yaav ya Mbany, the office of *Sanama* or governor of the newly acquired lands to the south (between Sandoa and Dilolo). Mukanza died during a campaign against his Nkongo neighbors to the northwest and, according to the Carvalho and Duysters versions, was succeeded by the *Sanama* Yaav ya Mbany, grandson of his brother Mulaji. Yaav ya Mbany (or Yaav II) is said to have reigned at the turn of the nineteenth century. He reportedly pursued his predecessor's war against the Nkongo and initiated an eastward drive against the Luba Samba of Kayembe Mukulu which was completed only several years later. The conflicting versions of the next reign, that of Cikombe Yaav, suggest that his authority was seriously challenged in some parts of the empire. Although he is credited with a normal span of office in the recent missionary compilations quoted above, Duysters asserts that Cikombe was faced with the competing claims of his paternal uncle and forced to abdicate after two months in favor of Naweej ya Ditende, a grandson of the late Yaav II. As for Dias de Carvalho, who collected most of his information in the western, or trans-Kasai portion of the empire, he simply claims that "Quicomba" did not reign at all.

We are on firmer ground with Naweej ya Ditende (Naweej II), a ruler about whom there is a virtually unanimous consensus among our various sources. Naweej effectively repulsed the bid for the throne made by Cikombe's rival and, like his predecessors, conducted (without much success) military campaigns against his northern neighbors, the Nkongo and Kete. However, he decisively defeated Kabeya Ilunga and completed the conversion of the lands of Kayembe Mukulu into a vassal chiefdom guarding the eastern marches of the Lunda homeland against the Luba. It was also during his reign that the Cokwe first threatened the relative stability of the empire which they were to destroy some fifty years later. In their northeastward advance, they came into conflict with the Minungu tributaries of the Mwaant Yaav and although a Minungu chief killed the Cokwe leader Mwa Cisenge (Mwatshisenge), the Mwaant Yaav had to intervene, both diplomatically and militarily, to contain the Cokwe and to force their most advanced elements, who had reached the Kasai around 1840, to send tribute to the Musuumb.

It was under Naweej's reign that the first Portuguese traders reached Lunda directly. For several decades, Portuguese officials and businessmen had shown a growing interest (not exclusively centered on Lunda) in establishing new lines of contact with the interior of South Central Africa. On the governmental side, this had led to the dispatching of the Lacerda expedition which reached Kazembe from Tete in 1798, and to the transcontinental journey of two *pombeiros,* Anastacio José and Pedro João Baptista, who reached the Musuumb from Angola in 1805 and eventually made their way to Tete by way of Kazembe, returning to Angola in 1814. Concomitantly, two official Lunda embassies had traveled to Luanda in 1807-1808 and again in 1811.[29] These and similar attempts during the early years of the nineteenth century were actively discouraged by Kasanje, whose continued prosperity depended on its preserving a monopoly of access to Lunda, but new African intermediaries had already appeared on the scene and opened up new itineraries to the lands of the Mwaant Yaav. From Bihé across the upper Cuanza to the upper Kasai, a new route was pioneered by the Ovimbundu and soon involved the Cokwe and Lwena peoples. The trade channeled along this new itinerary flowed predominantly in the direction of Benguela, but alternative routes circumventing Kasanje (notably across the Songo) were also being explored from the hinterland of Luanda and several Portuguese traders eventually made their way to Lunda during the 1830s and 1840s.

The first documented account of such an expedition is that of Joaquin Rodrigues Graça, who reached the capital of Naweej Ditende on September 3, 1846, by way of Bihé and returned to Luanda in 1848.[30] Although Graça failed to accomplish his particular objectives at the Musuumb, direct access to Lunda had now been achieved and, in the same year, 1848, another trader by the name of Lourenço Bezerra Correia Pinto established himself on the right bank of the Kasai, whence he was called by the Mwaant Yaav to settle at the Musuumb in 1849 or 1850.

There is no doubt that Naweej II desired to open direct commercial relations with the Portuguese, although some of his trading methods must have appeared rather unorthodox even to seasoned backwoodsmen like Graça and Bezerra: he would seize all the goods of a trading caravan; then designate a number of villages from which the traders would be permitted to take all the slaves they needed in compensation. Yet this was probably the only way in which the king could reconcile his absolute authority with the bilateral (and thus necessarily equalitarian) nature of commercial relations: he would appropriate the goods as an exercise of royal prerogative, then grant compensation out of royal magnanimity.[31]

Naweej II, who had had to face three pretenders and several conspiracies during his lifetime (the last one shortly before Graça's visit in 1845), had grown into a cruel and probably paranoid man by the end of his life, and rumor has it that his cousin, Mulaji a Mbala (Mulaj a Namwan), hastened his death by smothering him and then appropriated the royal bracelet (*lukano*), in complete violation of traditional rules of succession.[32] Mulaji a Mbala, who was already an old man when he perpetrated his palace coup, apparently tried to placate his opponents by appointing Naweej's favorite nephew, Muteba ya Cikombe (who had helped defeat the 1845 conspiracy), as his *Nswaan Mulopw*, but soon accused him of being responsible for his illness and attempted to liquidate him, whereupon Muteba fled, taking with him one of Mulaji's wives. When Mulaji died in 1857, his son Cakesekene Naweej, who had been made *Nswaan Mulopw* after Muteba's disgrace, attempted to seize the throne but was ousted and killed by Muteba's supporters after a short civil war.

Muteba ya Cikombe (or Muteba II), one of the few Mwaant Yaav to have died a natural death during those troubled times, reigned until 1873. During his time, trade continued to expand, with Lourenço Bezerra now firmly established in the land, but the king was faced with the continued threat of Cokwe incursions and with the growing insubordination of the governor of the southern provinces, the *Sanama* Mbumba (Mbumb Muteb a Kat), who had been appointed to his post by the late Mulaji.[33] Mbumba's mother, who held the dignity of Rukonkesh, was killed at Muteba's behest and the king's own *Nswaan Mulopw* even defected to the *Sanama*'s side, but Mbumba bided his time until Muteba's death, when he mounted a conspiracy against the new Mwaant Yaav, Mbala II (Mbal a Kamong Isot), a descendant of

Naweej II. Mbala II tried to flee to Kayembe Mukulu but was killed by Mbumba's followers in 1874.

Mbumba's accession marks the beginning of a series of violent convulsions which culminated with the Cokwe invasions of 1885-1888 and with the complete collapse of the Lunda empire, but it also coincides with the beginnings of European exploratory interest in the area, signaled by the expeditions of Vernon L. Cameron—who never reached the capital—and of Dr. Pogge, who reached it in 1875, to be followed by several others during the next decade. The reign of Mbumba was, in effect, a protracted war of succession[34] in which the Cokwe took an increasingly active part. Like many of his predecessors, Mbumba, who as *Sanama* had been in charge of guarding the empire's southern border against the Cokwe, hired some of them as mercenaries and used them in his bid for the throne. After his accession, he relied on Cokwe bodyguards and, in effect, came to depend upon Cokwe support. Even though he liquidated his predecessor's wife and children,[35] the line of Naweej II (descended from Yaav Naweej, or Naweej I) was not extinct, and Mbumba was finally defeated and killed by another son of Naweej II, Cimbindu a Kasang. Before he died, Mbumba had aparently come to realize the alarming extent of Cokwe infiltration into the affairs of the state, but he was unable to stem it and his followers would prove to be even more powerless.

From their original habitat near the headwaters of the Kwango, Kwilu, and Kasai rivers, the seminomadic Cokwe had been engaged in the not unrelated activities of hunting, collection and trading of ivory and wax, gun-running, and occasional piracy. In the process, they had been extending their control over an ever-increasing area. By the mid-1860s advanced parties had penetrated deep into Kasai near the area of modern Kananga (Luluabourg), and the Cokwe were on the march in several directions, mostly northward between the Kasai and Kwilu areas. Within little more than a decade, they virtually controlled a wide corridor extending as far north as Tshikapa and thus effectively cut off the Mwaant Yaav from the westernmost portions of his empire—an area over which his control had, in any case, become somewhat less than firm.

The Cokwe expansion into the Lunda homeland proper began in the 1870s and was brought about in part, as we have seen, by dynastic struggles among the members of the Lunda aristocracy. At approximately the same time, what little remained of the Mwaant

Yaav's control over the Kazembe domains to the east was being shattered by the ruthless empire-building activities of Msiri and his Nyamwezi (Bayeke) followers, who, by the time of Mbumba's accession to the throne, controlled the whole expanse between the Lualaba and Luapula rivers. Msiri's trade connections, of course, were almost exclusively oriented toward the Indian Ocean coast, while the commercial chain of which the Cokwe held one end reached all the way back to Benguela (and hence across the Atlantic to Europe and the Americas). Thus the Lunda (and, for that matter, the Luba) found themselves seized between the advancing tentacles of two giant exploitative systems reaching at them from opposite ends of the globe. Against such odds, the old Lunda empire did not stand a chance.

After Mbumba's death in 1883, his murderer and successor, Cimbindu, ruled for only five months before he was overthrown by Kangapu Naweej. Kangapu himself was killed a year later by Mudiba, yet another son of Naweej II. Meanwhile, the Cokwe were pouring into the Lunda homeland across the Kasai. Mudiba tried to stem the invasion but was killed at the battle of Mueji in October 1885. The disaster was complete: the *Nswaan Muruund* and several other dignitaries were taken prisoner, the capital was sacked, and some six thousand Lunda were taken into slavery. While these disastrous developments were taking place in Lunda, and without the knowledge of its inhabitants, statesmen gathered in Berlin had partitioned a land which only a handful of white men had even seen, and, within a few years, the beleaguered Lunda homeland discovered that it had acquired a new master: the Congo Free State.

A new king, Mukanza (Mutand Mukaz), was invited to take the throne in the land of Kayembe Mukulu, where he and other notables had retreated, but in January 1887 a fresh invasion began, and this time the Cokwe tide engulfed the Nkalaany Valley, the cradle of Lundahood, and three thousand more Lunda were taken slaves. Mukanza fled ingloriously and was deposed in favor of Mbala III (Mbal a Kalong) who reoccupied the Nkalaany Valley; but, before the year was out, Mbala fell at the hands of the two sons of the late Mwaant Yaav Mbumba. Like their father before them, the two brothers, Mushidi and Kawele, made use of Cokwe troops in their bid for power, but the Cokwe were now the real masters in the land. In 1888, they took the capital (then at Kawend), reduced the population to slavery, and forced the new Mwaant Yaav, Mushidi, to seek refuge first among the Luba and then on the edges of Kaniok territory. Cokwe occupation of the Lunda homeland lasted a full decade, during which the spirit of resistance was kept alive by Mushidi, by his brother Kawele, whose unflagging courage made him a sort of national hero of the Lunda, and by local chiefs, such as Mwene Kapanga and Mwene Mpanda. By 1898 the Lunda felt strong enough to counterattack and began to drive out the invaders, who had neglected to fortify their settlements. They won a first battle against the Cokwe leader Mawoka on the upper Lulua, but a Portuguese detachment followed by troops of the Congo Free State under Commander De Clerck intervened to put an end to the fighting.

The agents of the Congo Independent State, having had to face the Cokwe for several years either as suppliers of rifles to recalcitrant chiefs or in actual skirmishes where they came to the assistance of their African trading partners, were initially inclined to favor Mushidi,[36] but the king had no reason to welcome the *pax Leopoldiana* if it prevented him from recovering his lost territories. Mushidi soon ran afoul of the new authorities and took to the bush, where he and Kawele kept on fighting the Europeans until 1910. Meanwhile, Chief Kapanga, who had also been a major artisan of the Lunda recovery and who had brought up the son of the late Muteba ya Cikomb during the wars of succession, had made his peace with the newcomers and his protégé was eventually recognized by the Free State in 1907 as Mwaant Yaav Muteba, although he was not ritually invested until 1916.[37]

Muteba now consolidated his power with the help of the Belgians, and the two brothers, betrayed by several Lunda notables, were finally put to death on his orders. But the throne that Muteba had gained was not what it used to be. Not only were its eastern and western provinces irretrievably lost — and the loss itself sanctioned by the new boundaries negotiated by the European powers — but the southern half of the homeland was dotted with Cokwe villages as far east as Kayembe Mukulu, and the invaders actually outnumbered the local population in most areas. Even where Lunda chiefs had succeeded in maintaining or reestablishing themselves, the tributary network had collapsed, and the Belgians would not have tolerated its forcible restoration, even if the Mwaant Yaav had possessed the military means to do so. Thus, in its reduced state, the moribund Lunda empire entered the colonial age.

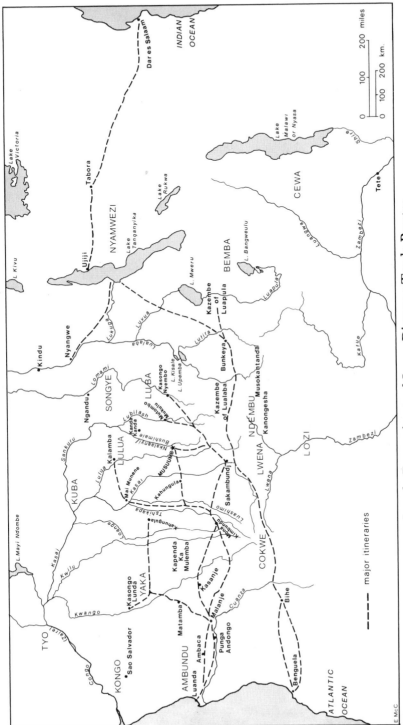

Lunda Expansion and Long-Distance Trade Routes

2

The European Penetration and
Partition of Lunda

As with so many of the states in the interior of Africa, the fame of
Lunda grew during its age of expansion and came to the ears of
Europe after considerable delay. By the time Western merchants,
missionaries, and soldiers reached it, the power of Lunda had passed
its peak. Indeed, in many ways, the commercial appetites aroused by
the original reports were indirectly responsible, at least in part, for the
kingdom's decline.

Lunda's first major expansionist drive—the migration under
Cinguud and Kasanje—had the effect of putting it into contact with
agents of a Western power and introduced them to some Western
goods, so that it is not really accurate to talk of a Western discovery of
Lunda. Rather, advanced parties of the Lunda culture deliberately
sought to come into contact with the white men, and we are told (at
least if we adopt the prevailing version and chronology) that it was
upon learning about the Portuguese and their wares (presumably from
the Ambundu) that the migrants desired to make contact with the
foreigners. After the emergence of the Imbangala nation and the
founding of the Kingdom of Kasanje, the descendants of the Lunda
migrants settled to a life of trade with the Portuguese and jealously
tried to preserve for themselves a monopoly of access to the trade resources of the Kasai region.

Kasanje, of course, was only one of several trading states which developed in the hinterland of Luanda (and later of Benguela), and its
emergence concerns the Lunda homeland only to the extent that for a
long time it drained whatever goods Lunda had to offer in exchange

for Western commodities, primarily cloth and guns. There was little or no attempt during that period to establish any kind of Portuguese control over such remote areas as Lunda. Indeed, as demonstrated by the vicissitudes they experienced between the time of the Jaga invasions of the 1560s and the restoration of Portuguese primacy over the Mbundu kingdoms during the second half of the seventeenth century, following the Dutch interlude at Luanda (1641-1648), Portuguese resources were sorely taxed in the mere attempt to maintain control over the immediate hinterland and all they were able to do in respect to the trading states of the Kwango region, such as Matamba and Kasanje, was to secure an acceptable *modus vivendi*.

By the turn of the eighteenth century, a relatively stable pattern of trade had emerged in northern Angola, with Kasanje at its center, and the Lunda empire became the main purveyor of the Luanda trade, later countributing to the growing trade of Benguela and possibly of Loango.[1] What Lunda had to offer in those days was mostly slaves and ivory[2] and what they got in return was cloth, some tobacco, and a few manufactured articles, especially guns and gunpowder. The growth of the Lunda trade seems to have coincided with the territorial expansion of the empire and there may well have been some sort of self-perpetuating process involved: as the Lunda conquered new lands, more captives were funneled into the channels of the Atlantic slave trade while more weapons were made available to the Lunda, thus enabling them to expand the scope of their conquests. We do not know with any degree of accuracy, of course, just how extensive this trade may have been but we do know that, by the early seventeenth century, guns were common enough in the Lunda homeland for them to have been employed by Kanyembo and his followers when they carved out for themselves the kingdom of Kazembe.

No real penetration on the part of the Portuguese accompanied these developments. This was in part a reflection of the *modus vivendi* that had gradually been achieved with the trading chiefs of the hinterland. These chiefs were suspicious of any attempt by the Portuguese to acquire their slaves directly. Accordingly, trade was normally conducted through black or mulatto agents, the *pombeiros*.[3] Sporadic attempts by individual governors to deflect a portion of the trade to their benefit through the dispatching of Portuguese detachments were bitterly resented not only by the trading chiefs but also by the Portuguese firms themselves, which viewed this practice as unfair competition and as a danger to established trading patterns. Thus in 1703 the municipal council of Luanda ordered all white men and mulattoes to return to the coast, and in 1721 a law enjoined the governors from taking part in the slave trade.[4] Such prohibitions could hardly be expected to be entirely effective, of course, and after several reiterations, the ban on trading in the hinterland by white men was eventually lifted, in 1758. Even so, the trading states of Matamba and Kasanje were sufficiently strong to interpose an effective barrier against any direct contact between the Portuguese and Lunda until the process of Lunda expansion into the Kwango had built up a tremendous pressure against these intermediaries. Although Kasanje and Matamba managed to maintain themselves, their position was weakened as a result of a number of factors. They had to accept the permanent presence of Portuguese traders (in the case of Kasanje, this had become a reality by 1797).[5] More important, the Lunda trade began to flow into new channels, which had the effect of circumventing Kasanje's monopoly. Chief among these were the routes opened by the Ovimbundu from Bihé to supply the port of Benguela. One such route followed approximately the course later adopted by the Benguela railway and entered the Lunda homeland in the area of modern Dilolo, while the other took a more northerly path to Mona Kimbundu (or Quimbundo), between the upper Loange and the upper Tshikapa, from which point other routes led either to the Musuumb of the Lunda sovereign or (by following the Tshikapa downstream to its confluence with the Kasai) to Mai Munene and the western Luba lands.

Eventually, traders of the Luanda hinterland outflanked Kasanje on the south and reached Mona Kimbundu, whence they traveled along the same routes as the people from Bihé. This alternative route from Luanda by way of Malanje and Mona Kimbundu came to be known in the nineteenth century as the Caminho Grande, the Caminho de Quimbundo, or the Caminho de Munene Quissesso, after the nickname given by the Africans to Saturnino de Sousa Machado, one of the two enterprising brothers who were responsible for opening this route to direct Portuguese trade around 1850 and who established the first permanent European trading agency at the court of the Mwaant Yaav, under Lourenço Bezerra.[6] In addition to its considerable commercial importance, the Camhinho Grande played a major role in the systematic exploration of Lunda a quarter of a century later, being the route followed in whole or in part by the expeditions of Pogge and Lux (1875), Otto Schütt (1877), and Max Büchner (1878-1881).

Here again, of course, the term "exploration" must be qualified. When Pogge, Schütt, or Büchner traveled into Lunda, they were not really "discovering" anything that had not been known for decades, not only to pombeiros but also to Portuguese traders or adventurers. At the same time, however, there had been no serious attempt by the Portuguese authorities to extend any sort of political authority over Lunda. This was in keeping with the policies of other European powers trading on the Atlantic coast of Africa. In point of fact, the Portuguese had gone to greater lengths than any others to subjugate or destroy the states of the immediate hinterland, but such expeditions were all related directly or indirectly to commercial preoccupations rather than to the late nineteenth-century concern with territorial expansion — which the Portuguese did not generate in any case. Not only was there no question of trade following the flag in Angola, but also, for a considerable length of time, the flag was extremely reluctant to follow the trade. In 1765, for example, the governor of Luanda, Dom Francisco de Souza Coutinho, was of the opinion that giving military protection to Portuguese traders would only encourage them to impose unfair conditions on their African partners and thus jeopardize the normal flow of trade.[7]

At the same time, however, there had been a recurrent preoccupation on the part of Portuguese official circles with the establishment of overland links between their possessions on the Atlantic and Indian Ocean coasts. The project was first studied in 1590; it was sporadically resurrected during the following two hundred years until official proddings from Lisbon eventually produced two related but uncoordinated attempts to open a transcontinental route. The expedition which left from Tete in 1798 under Francisco José de Lacerda e Almeida went no farther than the capital of the Kazembe, where Lacerda died and where in any case he had probably been preceded by private traders.[8]

From the Angola side, on the other hand, attempts to establish a link with Mozambique were related, at least in part, to the recurrent preoccupation with finding an access to Lunda that would not be under Imbangala control. Franciso Honorato da Costa, a Portuguese merchant who had suffered financial reverses and become the official Portuguese commercial representative in Kasanje, had been exploring the possibility of opening new trade routes across Songo country when he was invited by the governor of Luanda, Miguel Antonio de Melo, to collect information on possible transcontinental itineraries. This seems to have prompted Honorato to dispatch two of his agents on an exploration of Songo country in November 1802, but the two pombeiros were detained by a local chief. In response to further proddings from de Melo's successor, Fernando Antonio Soares de Noronha, Honorato secured the release of his emissaries in November 1804 and the two men, Anastacio José and Pedro João Baptista, proceeded to the Musuumb (which they reached in 1805) then, as related earlier, to Kazembe and Tete, returning to Angola in June 1814.[9] In 1815, an attempt to capitalize on the relative success of this journey by organizing a regular service of transcontinental couriers failed to materialize, but in the same year that the two pombeiros had returned to Angola, a Portuguese trader by the name of João Vicente da Cruz first established himself in Lunda territory.[10] It would seem likely that during the following years, *sertanejos* continued to infiltrate Lunda in their quiet, slow way so that when Joaquim Rodrigues Graça reached the capital of Naweeji II in 1846, he recorded without any apparent surprise that he had been preceded there by a European trader named Romão. African traders, of course, had long been familiar with the Mwaant Yaav's court, and Graça's mission, which had been to establish direct trade relations with the Lunda potentate and thus to break the monopoly of the Kasanje route, was partly foiled by the arrival of a caravan led by four pombeiros in the employ of Ana Joaquina dos Santos Silva (*alias* Dembo ya Lala), a Ndongo tradeswoman for whom Graça had worked at one time.[11] Thus, by the late 1840s, the 200-year monopoly of Kasanje over the Lunda trade was being effectively undermined.

Kasanje's position was being challenged on several fronts. First of all, of course, came the formidable competition offered by the Ovimbundu caravans from Bihé who had been trading on the Lwena marches of Lunda ever since the turn of the nineteenth century, had established contacts with the Lozi in the 1830s, and were now reaching into the Lunda homeland. On Kasanje's western flank, the Portuguese of the Luanda hinterland had been seeking to circumvent the Imbangala ever since the beginning of the century and their growing impatience with the Jaga finally erupted in the form of two military expeditions between 1848 and 1852 which further weakened Kasanje's position. Although the Imbangala kingdom soon recovered its military strength and soundly defeated a Portuguese force in 1862, thus ensuring its independent survival until 1910, the crisis led to the deliberate build-up of Malanje by the Portuguese as an alternative marketplace.[12] The growth of Malanje after 1850 was assured by the opening of the Caminho Grande under the auspices of the Machado brothers, with an anchor at the Lunda royal court in the person of Lourenço Bezerra. Thus, while Kasanje retained the military strength to interdict the shortest route into Lunda, it was powerless to prevent the development of alternative accessess and the monopoly they guarded gradually lost its commercial value.[13]

Of even greater importance, in the long run, was the gradual decline of the slave trade. Despite occasional efforts by Portuguese governors such as Souza Coutinho to introduce a plantation economy, Angola at the turn of the nineteenth century depended almost exclusively on slave-dealing. After soaring until the mid-seventeenth century, the slave trade had slumped somewhat as a result of political vicissitudes but, by the 1750s, annual slave exports from Luanda and Benguela again exceeded 10,000, as they had done until the capture of Luanda by the Dutch. In 1765 — a "good" year — official shipments from the two ports amounted to 17,247 slaves,[14] and in 1789, 88.1 percent of the colony's income was derived from duties collected on the export of slaves.[15] In 1834, however, Portugal yielded to international pressure and outlawed the slave trade (though not slavery itself). Although it took many years for the ban to be effectively enforced, the whole economy of Angola entered into a protracted period of decline and reconversion from which it did not emerge until some fifty years later.

By 1845, naval surveillance had virtually closed the trans-Atlantic trade from Angola, whose total exports had declined in value to a mere fifth of what they had been only twenty years earlier. In that same year, after nearly three centuries of Portuguese activity, the total white population of Angola was reckoned at only 1,832. A decree of 1858 made it illegal to reduce any person to slavery, and finally, in 1875, all remaining slaves were formally set free. To be sure, the Portuguese soon reintroduced other coercive measures designed to ensure a continuing supply of cheap African labor — a policy that was prompted by the development of a plantation economy and culminated in the introduction of the forced labor system in 1899 — but the old slave trade was dead and, with it, a set of economic patterns which had had deep repercussions on the growth of Lunda. The changing nature of trade after 1840 was to produce far more serious effects on that Central African kingdom.

In 1834, the same year which saw the outlawing of the slave trade, the royal monopoly on ivory was also abolished. The intended effect of this measure was to stop the contraband flow of ivory to non-Portuguese establishments and to regain by way of export duties some of the income which the monopoly system had patently failed to secure for the Portuguese treasury. Price levels trebled almost immediately, and ivory exports from Luanda jumped from 3,000 pounds in 1832 to 105,000 in 1844, by which time ivory had become the territory's most important export. Exports of wax from Angola became almost equally important during the same period: shipments from Luanda alone rose from 52,690 pounds in 1844 to 1,698,248 pounds in 1857.[16]

No single tribe in Angola was better prepared to take advantage of these new conditions than the Cokwe. They were excellent hunters and occupied one of the major wax-producing areas in the land. Also, their traditional socio-economic structures were particularly well suited to the combined activities of hunting and collecting. Until the boom in ivory and wax, Cokwe involvement in long-distance trade patterns had been marginal. They had sold small quantities of wax, ivory, and slaves, but, since the latter commodity had dominated the market so completely and the Cokwe did not as a rule sell their own people into slavery, the benefits they had derived from the coastal trade had been limited. Guns were the only Western exports in which the Cokwe, as a hunting people, were really interested. Being competent blacksmiths, they were able to maintain them and thus to keep them in service longer than many of their neighbors. Thus, despite their own limited exports, they had been able to accumulate a respectable quantity of firearms by the first part of the nineteenth century. The artificially depressed prices for ivory caused by the monopoly system restrained the growth of large-scale elephant hunting by the Cokwe until the late 1830s. On the other hand, their strategic location on routes which the Ovimbundu were trying to develop from Bihé to Lunda led them naturally to exploit their nuisance potential by levying duties (*milonga*) on passing Ovimbundu caravans. Of course, the benefits of this sort of economic parasitism were limited, literally, to what the traffic would bear, and traders from Bihé soon learned to avoid Cokwe territory. But the dramatic rise of the trade in wax and ivory released the Cokwe from their relative stagnation and launched them on a spectacular course of commercial and political expansion. The production of wax, one of the two staples of the Cokwe export economy, was by nature a relatively parochial enterprise which could be carried out by settled elements of the population; and for which the weight of transportation costs was to discourage for a long time the prospecting of remote areas. With ivory, on the other hand, prices were sufficiently attractive to warrant long-distance expeditions and, in any case, the Cokwe themselves had virtually exterminated the elephants from their own land by the mid-1850s, thus making it imperative for them to follow the herds in their gradual retreat to the north.

This process can be followed geographically almost step by step. The traditional Cokwe heartland in the early part of the nineteenth century had been situated in the vicinity of the eleventh parallel south, and in 1850 the Cokwe were still found to the south of the tenth parallel. By 1850, the southernmost limit of elephant-hunting grounds had receded to the ninth parallel, and it continued to retreat toward the equator at the approximate rate of one degree latitude per decade.[17] By the late 1850s, advanced groups of Cokwe were found in the vicinity of the ninth parallel, and by 1866 they had advanced to the eighth parallel.[18] But even at that point a few adventurous hunters had crossed the Kasai near Mai Munene and had advanced to the Lulua where they introduced firearms to the western Luba — specifically to an ambitious local chieftain named Mukenge Kalamba, who used them to forge his followers into a new political entity, the Bena Lulua — a name coined by the Cokwe.[19] Within a few years, Kalamba had developed a thriving market where the Cokwe had a virtual monopoly. Since Kalamba's people supplied the market with ivory, rubber, and slaves, the Cokwe turned more exclusively to trade, and, when Schütt and Büchner traveled into the area some ten years later, they found the Cokwe in control of the caravan traffic to Mai Munene and beyond. Elsewhere, however, notably in the Lunda homeland, the Cokwe offered their services as hunters and shared the ivory with the local chiefs, an arrangement which for a long time satisfied both parties and which the Mwaant Yaav's decision to exact tribute in the form of ivory after 1858 made almost mandatory for subordinate Lunda chiefs.

As the elephant herds retreated toward the equator and ivory became scarcer, the demand for native rubber, which began around 1870, offered the Cokwe a new trading opportunity. As with wax and ivory, the Cokwe had originally collected rubber in the forests of their original homeland of Quiboco but eventually followed the trail of native rubber along the forested banks of the northward-flowing tributaries of the Kasai, which further served to reinforce the south-north orientation of their trade and travels. Here again the Cokwe found that rubber flowed naturally into Kalamba's market from local suppliers of the Lulua region but that, in Lunda, the local population did not care to collect it[20] and was willing to let the Cokwe settle in the forest areas which they themselves did not inhabit, preferring to build their villages in open country.[21]

The transformation of the Cokwe from a nation given to hunting, gathering, and trading to one of migrants and conquerors seems to have been gradual and only partly deliberate. More than by any other factor, it was produced by the repercussion of economic changes on the Cokwe social structure. Unlike the Lunda who exported their institutions and, to a certain extent, their political elite, but who did not migrate themselves in large numbers, the Cokwe after 1840 were a nation on the march, but a nation which did not subjugate so much as it integrated other people. In the words of Joseph Miller, "The Cokwe were, in fact, able not only to integrate other people up to several times their own numbers but also to convert them to Cokwe ways of living, literally making Cokwe out of their Lunda, Luchaze, Pende and Lulua neighbors."[22] The most important single factor in this process was the "pawnship" system which permitted "the transfer of rights over an individual from one lineage to another" and was widely used as compensation for damages. What made the system crucial for Cokwe expansion was the fact that it could be applied to non-Cokwe women. "Pawnship provided a status within Cokwe society which was independent of membership in a Cokwe lineage" but which "automatically bestowed full lineage membership on the pawn's children by a Cokwe husband."[23] Under such a system, the acquisition and exchange of female pawns became primarily an economic phenomenon and women presented very much the same characteristics from the viewpoint of capital accumulation as did cattle in the pastoral cultures of eastern Africa — divisibility, mobility, and natural increase.[24] Indeed, from a strictly economic viewpoint, there was little difference between pawns and slaves and it is not surprising therefore to find many early Western observers referring to Cokwe women "slaves." From a socio-demographic viewpoint, however, since the children of such women by a Cokwe father were brought up as full members of the lineage, the effects of the system permitted the society to experience tremendous population growth without excessive strain on the social structure. The Cokwe probably began acquiring alien women in this fashion at about the same time that they acquired their first firearms and, as they expanded the scope of their trading and raiding, the procurement of women (through trade, at Kalamba's market, or through raiding in the Lunda homeland) remained a dominant part of their activities. Population pressures must almost certainly have been felt in the small Cokwe heartland of Quiboco as early as the 1840s and gradual migration by entire villages began to take place in opposite directions.[25] The less important of these two currents, the southern, is not very well documented and is not, in any case, of direct relevance to us. To the north and northeast, however, Cokwe migrants naturally followed the trails blazed by their own hunting and trading parties. Cokwe military tactics depended largely on mobility and superior firepower; they had no use for fortifications. But long before an actual military confrontation took place, the Cokwe had infiltrated their neighbors' territory unopposed — both ecologically, by settling in the unoccupied forest areas, as among the Lunda, and economically, by performing functions such as those of hunters and blacksmiths which were left vacant by their hosts.[26] During the first stage, they would readily accept the authority of the local chief and even, in the case of the Lunda empire, acknowledge that they owed tribute to the Mwaant Yaav, although it is not clear how frequently, if ever, they paid it. According to a 70-year-old Cokwe informant interviewed by Dias de Carvalho, Naweej II had granted a concession to Cokwe hunters in the Lunda heartland in the early 1840s,[27] but when the Cokwe attempted to establish a chiefdom at the point where the trade routes from Kasanje and from Bihé converged, there was a brief conflict ending with the creation of a Lunda military march under Mona Kimbundu which soon became a major turntable for the trade into the Lunda and Lulua regions. Within a generation, however, the Cokwe had completely infiltrated the area and their migratory tide was advancing northeastward toward the Kasai and in a northerly direction toward Mai Munene, with the Tshikapa as their axis. Eventually, this northward migration covered the entire area between the Kasai and Kwilu rivers as far north as Mwata Kumbana and Mai Munene with a number of forays toward the Kwango and Sankuru rivers. This process continued up to the time when the forces of the Congo Free State consolidated their control over the area.

In the Lunda homeland, as indicated earlier, Cokwe expansion really began with their intervention in the Lunda wars of succession. By the mid-1850s, Mona Kimbundu's outpost had been swamped by Cokwe migrants and Mbumba was appointed *Sanama* by the successor of Neweej II, Mulaji II a Mbala, for the purpose of defending the western bank of the Kasai and of keeping control over the southern trade route into the Lunda homeland. Instead, Mbumba reached a *modus vivendi* with the Cokwe, diverted their expansion to the north of his district in the direction of Mwasanza and Mataba and, upon the death of Muteba ya Cikomb in 1873, recruited Cokwe followers to back up his claim to the throne. Muteba's successor, Mbala II, was forced to flee[28] in 1874, and Mbumba became Mwaant Yaav. Although a number of Cokwe retainers remained at Mbumba's court, the two chiefs who had supported his bid for power withdrew to the west of the Kasai into an area which Mbumba had given them permission to plunder in reward for their services.

At that point, it appears that the Cokwe were already finding the Lunda homeland less attractive than the middle Kasai region. The Lunda tributary system, which had made vast quantities of ivory available for sale at the royal capital, was in almost complete disarray (in part because of the Cokwe themselves), and the traders found it more expedient to deal directly with local chiefs (who, in turn, often found it convenient to explain their failure to send tribute by Cokwe interference) or, increasingly, with the vastly more productive areas from which the Bena Lulua drew their supplies. But when it was no longer able to drain the resources of the far-flung empire, the Lunda heartland itself had little to offer: the elephants had been driven out or exterminated; native rubber (which the Cokwe collected themselves in any case) was not as plentiful as in the forest galleries of Kasai, and what mineral wealth there was to the east had fallen under Msiri's control. In fact, the only resource that was left to attract the Cokwe raiders was the population itself, which they captured by the hundreds as slaves or as pawns.

The vicissitudes suffered by the Lunda heartland after 1875 have been outlined in the preceding chapter, and there is no need to repeat them here. What is worth noting, however, is the fact that all contenders for the throne during this "time of troubles" (as well as a number of local notables) made use of Cokwe troops at one time or another. The Cokwe naturally made good use of the prevailing anarchy to roam across the Lunda homeland which, compared with some of the areas farther north,[29] must have seemed to them a point of less resistance. Even so, it was not until some ten years after their first intervention in the dynastic politics of Lunda that the Cokwe migrants began to pour into the Lunda homeland. Within three years this new wave of invaders had utterly destroyed the core of the old empire and emptied it of most of its population, but the land they had conquered was of little immediate value to them. The Lunda heartland had once stood as the center of a politico-economic network which no longer existed. Its own intrinsic resources were negligible. The vital arteries of the Cokwe exploitative network ran farther to the west and north and, to the newcomers, the Lunda homeland was in a sense a dead end. Cokwe villages were scattered thinly over the land, and no effort was made to organize the conquered area politically. The Cokwe had been able to assimilate individuals through the pawnship system, but, unlike the Lunda during their period of expansion, they were unable to integrate entire communities. Thus, when Mushidi and Kawele (or Zama, as he had been nicknamed) regained the initiative after 1896, there were few institutional remains of the Cokwe invasion, and those Cokwe groups that remained in the land—a fairly large number—found it easy to accept a *modus vivendi* which in most cases involved at least the nominal suzerainty of Lunda chiefs.

The year 1875, which marked the beginning of the agony of the Lunda empire, also witnessed the first signs of a new form of European interest in that part of Africa. Portuguese activities since the sixteenth century had ignited a slow-burning fuse which reached its end with the Cokwe destruction of the Lunda political structure; but the series of explorations that began with Paul Pogge's visit to the Lunda capital in 1875 were very much a part of a nineteenth-century pattern and heralded an entirely different kind of penetration. Pogge and Lux had followed the Caminho Grande to the Musuumb and relied on the good offices of the Machado brothers in their expedition, but when Otto Schütt (whose expedition, like Pogge's, had been organized by the

Berlin Geographical Society) reached Mona Kimbundu by the same route three years later, he found the trading caravans traveling north and followed their trail to Mai Munene; he was then prevented under various pretexts from pursuing his journey to the Mwaant Yaav's capital and made his way back to Luanda.[30] While he was returning to the coast, yet a third German expedition, cosponsored by the Berlin Geographical Society and by the German section of the newly organized International African Association, was embarking on the Malanje-Kimbundu route on its way to Lunda. Its leader, Max Büchner, made his way to the Musuumb despite Cokwe interference, but the Mwaant Yaav persuaded him not to pursue his journey to Nyangwe (a trip that V. L. Cameron had made in the opposite direction in 1873), while Lunda tax collectors effectively deterred him from returning to the Atlantic coast by way of the Kasai to check out Stanley's erroneous assumption that this river emptied into the Congo near the site of modern Mbandaka. Shortly thereafter, Von Mechow's ill-fated expedition attempted to explore the lower course of the Kwango by boat.

The number and frequency of German expeditions toward the middle Kasai during that time should cause no great surprise. Even though Bismarck was, as he put it, "kein Kolonialmensch" and remained cool toward the idea of overseas expansion almost to the eve of the Berlin Conference, there were enough German circles interested in a colonial empire to sponsor the kind of exploratory spadework which might open the way for subsequent territorial claims, should the government become converted to their views. King Leopold of Belgium, who was also interested in providing a scientific and humanitarian facade for his imperial ambitions, offered these circles a convenient vehicle with the creation at the 1876 Brussels Geographical Conference of the nongovernmental International African Association.

The autonomous national sections which were created after the Brussels conference soon became either moribund or exclusively preoccupied with the promotion of national interest. This was of course the case with the Belgian section (although Leopold took full advantage of his position as chairman of the I.A.A. to maintain as long as possible the fiction that his endeavors were undertaken on behalf of the international community), but also, to a lesser extent, with the French and German sections. The results of Schütt's journey to Mai Munene, as well as those of Silva Porto's venture into the land of

the Bashilange (Bena Lulua), were known to Paul Pogge when he returned to Central Africa for the second time in 1881, accompanied by
a young lieutenant, Herman Wissmann, who was later to become
governor of German East Africa.[31] Having made use, like all their
predecessors, of guides and bearers recruited for them by the Machado
brothers, Pogge and Wissmann traveled from Malanje to Mona
Kimbundu, but, after learning of the unfavorable circumstances
which then prevailed at Mbumba's court, they decided to travel north
instead of east and followed the route opened some fifteen years earlier
by Cokwe traders to Kalamba Mukenge's capital and thence to
Nyangwe, where Pogge and Wissmann parted ways. From Nyangwe,
Wissmann followed the Arab trade route across Lake Tanganyika to
Tabora and Zanzibar, while Pogge returned to the Lulua River, where
he founded a modest station near Kalamba's village. Pogge then
made his way back to Luanda, where he died on March 17, 1884, but,
on his way to the coast, the ailing explorer had met Wissmann, who
had sailed back to Europe from the East African coast and had returned once more to Angola, this time to set up a post on the Lulua on
behalf of the International African Association. Wissmann traveled
once more along the now familiar route with an escort raised at Malanje in November 1884, and he reoccupied Pogge's ramshackle
station, which received the name of Luluaburg.[32]

A few months earlier, two other German explorers, Böhm and
Reichard, had become the first Europeans to visit Msiri's capital at
Bunkeya. The aims of these German expeditions as well as of
Germany's official policy up to the time of the Berlin Conference were
somewhat ambiguous. On the one hand, the expeditions were officially undertaken on behalf of the International Association, which had
clearly become identified with Leopold's personal ambitions in Central
Africa; but, on the other hand, interest groups other than those
associated with Leopold could easily avail themselves of the humanitarian smokescreen which the Belgian monarch had found so useful
and which he could hardly afford to repudiate. Böhm and Reichard
had worked in close cooperation with Leopold's agents on the western
shore of Lake Tanganyika, but Wissmann and his German associates,
von François, Wolff, and Müller, had acted on their own and certainly
did not consider their expedition incompatible with the pursuit of
German interests.[33] As for Leopold, ever since he had hired the services of Henry Morton Stanley and created the Comité d'Etudes du

Haut-Congo in 1878, he had certainly shown little interest in the upper Kasai and had instead directed the energies of his agents toward the establishment of stations between the lower Congo and the great lakes, and toward the containment of French ambitions. In a letter of August 8, 1884, addressed to Bismarck, Leopold gave a summary description of the boundaries of the territory for which he was seeking international recognition; the southern limit he indicated between the Kwango and the Lualaba followed the 5°30' south latitude line, thus leaving out most of Luba land and all of Lunda.[34] Three months later, just before the opening of the Berlin Conference, Germany recognized the rights of the International Congo Association over a territory whose southern limit was the 6° south parallel. In fact, however, Bismarck had never had any real designs on the Kasai, and, of all the powers involved in the scramble for Africa, Germany was certainly the most favorable to the creation of the Free State, if only because the latter was expected to act as a buffer against French and British appetites.[35] On the other hand, the Free State itself seems to have attached little importance to the area in question, at least initially. In the declaration of neutrality to the powers of August 1, 1885, the sixth parallel was once more referred to as the southern limit of the Congo between the Kwango and the Lubilash rivers. Again, the whole of Lunda was thus being left out of the Free State's boundaries.

If Germany and even, to a certain extent, the Free State were not particularly concerned about Lunda, Portugal, for its part, had been taking decisive steps to substantiate what it considered its "historical rights" over the area. Capelo and Ivens, who had already explored the upper Kwango in 1878, visited Msiri at Bunkeya in late 1884 during their journey from Moçamedes to Quelimane.[36] But in Lunda itself no one championed Portugal's cause more persistently than Major Henrique Dias de Carvalho. Carvalho left the vicinity of Malanje in October 1884 (i.e., at the time when Bismarck was sending out invitations to the Berlin Conference) and proceeded to set up stations on the upper Kwango, then in the westernmost reaches of the tottering Lunda empire. There, in the lands of Cassassa (a *cilool* of Mwene Kapanga) on the upper Kwilu, he met one of the numerous pretenders to the Lunda throne, a descendant of Naweej II by the name of Tshibwinza Samadiamba, who requested his aid to travel back to the Lunda capital across Cokwe-held territory. Carvalho took the pretender along and continued his journey toward the northeast to the seat of Chief Kahungula Muteba, on the middle Lushiko. Kahungula Muteba, who had been virtually autonomous of the Mwaant Yaav for some time and who, like many of his peers, had reached a satisfactory accommodation with the Cokwe, recognized Portuguese sovereignty on October 31, 1885.[37]

In January 1886 Carvalho went on to cross the Tshikapa in the vicinity of the present border between Congo and Angola but was informed that he would not be permitted to cross the lands controlled by the Cokwe of Chief Cisenge as long as he was accompanied by the Lunda pretender. The expedition then traveled south between the Luachimo and Chiumbe rivers. There, on June 12, Carvalho, who was probably beginning to have some doubts regarding the political usefulness of Samadiamba, signed a protectorate treaty whereby the pretender recognized Portugal's authority over the whole of Lunda.[38] At the same time, however, Carvalho was negotiating with Cisenge, not only to make him accept Portuguese sovereignty (which he did on September 2, 1886), but also to restore peace between the Lunda and Cokwe so that the treaty signed earlier by the "Mwaant Yaav-elect" (as Carvalho affected to call him) could be turned into something more than a mutually agreeable fiction.[39] What Carvalho did not realize (or chose to ignore) was that Cisenge did not really have the power to stop Cokwe incursions into the Lunda homeland and that, in any case, all sides involved in the Lunda civil war were making use of Cokwe mercenaries to bolster their ambitions.[40]

Nevertheless, having successfully negotiated with Cisenge, Carvalho resumed his progress toward the Musuumb but the pretender, who apparently feared for his safety, preferred to remain behind. The Portuguese explorer then advanced to the Lunda capital, where, on January 8, 1887, he met Mukaza, who had seized the throne about a year earlier with the help of Cisenge's followers. Ten days later, having either despaired of Samadiamba's chances or simply decided that two treaties were better than one, Carvalho and Mukaza signed another protectorate treaty whereby Portuguese sovereignty over Lunda was again recognized.

If Carvalho had entertained any illusions regarding the extent of Mukaza's authority in the land, they must have been dispelled when, shortly thereafter, Cokwe bands supporting the candidacy of yet another claimant, Mushidi, laid siege to the capital and forced him to flee ingloriously. In May the inept Mukaza was deposed in favor of his

Nswaan Mulopw, Mbala, who tried to persuade Carvalho to remain at his court, but the Portuguese officer, who must have realized that the political situation was too precarious for his mission to accomplish any lasting results, decided to start on his return journey.[41] On his way back to Luanda he took with him some emissaries dispatched by Samadiamba, who was still waiting hopefully in the lands of Chief Kahungula of Mataba, and brought them to the acting governor of Angola in March 1888 in the hope of encouraging the Portuguese government to pick up the option opened by the Lunda treaty.

After studying the impressive results of the Carvalho expedition, the Portuguese government felt that it could advance its claims with greater assurance, especially in view of the fact that the Congo Free State had not achieved, as yet, any effective occupation south of the sixth parallel. Wissmann's station at Luluaburg had been taken over by the Free State in 1886, but all attempts to penetrate farther south had met with relative failure. Wissmann's attempt to navigate the Kasai above the river's confluence with the Lulua had been stopped at the falls, which the German officer named after himself.[42] DeMacar's efforts to penetrate the area situated to the southwest of Luluabourg had ended in near disaster due to Cokwe hostility.[43] Similarly, the attempts to ascend the Kwango in 1885 and 1886 — by Wolff and Büttner, and by Dr. Mense and Reverend Grenfell — had been blocked by the determined hostility of "Mwene Putu" Kasongo, *Kiamfu* of the Bayaka. As for the Lunda heartland, its occupation would not be attempted by the Free State for another ten years.

Thus, when in December 1888 Agostinho de Ornelas, a senior official in the Portuguese Foreign Ministry, suggested to the Belgian ambassador in Lisbon that the time might be ripe for the two powers to determine more precisely the limits of their respective possessions in Central Africa, he may have felt that Portugal would be negotiating from a position of relative strength. In fact, however, the Portuguese position was vulnerable on at least two counts. First, Portugal was anxious to secure the Free State's recognition of its "rights" over the area separating Angola from Mozambique, in view of the threat of British penetration northward from South Africa. France and Germany, who were only too glad to check Great Britain, had endorsed Portugal's claims as early as 1886, and a similar recognition by the Free State would have enhanced Lisbon's pretensions. Thus Portugal was approaching the Free State as a petitioner. Lisbon's position was further weakened by the fact that several maps published after the Berlin Conference (including even some Portuguese maps) as well as the above-mentioned agreements with France and Germany had not included Lunda in the Portuguese sphere of influence. The fact that it had equally been excluded from the limits of the Congo Free State, by that latter power's own definition of its boundaries, was hardly the sort of consideration to bother King Leopold. Answering that point some time later when the crisis had escalated to its maximum, the Belgian geographer A. J. Wauters (at that time an enthusiastic supporter of the Free State) simply stated: "If the maps and documents issued by the Congo Free State in 1885 left out the Mouata Yamvo's territories, it was only because the Free State was not at that time in a position to occupy it effectively." But, he went on to say, the situation had now changed with the creation of Luluabourg and the dispatching of expeditions toward the south.[44] For all its casuistry, Wauters' argument represented an accurate description of Leopold's policy.

In response to Ornelas's offer of a settlement, the Free State replied by demanding in exchange for its recognition of Portugal's "historical rights" a huge area extending from Lake Bangweulu to the western bank of Lake Nyasa (Malawi) between 11°40' south latitude and the southern border of German East Africa.[45] It seems highly doubtful that the Free State ever seriously believed that Portugal might accept such exorbitant terms and indeed, on February 24, 1890, the Belgian ambassador to Lisbon confirmed that Portugal was reluctant to negotiate on that basis. The probability that the offer from Brussels represented merely a dilatory maneuver would appear to be reinforced by the fact that, while it was drafting its reply to Lisbon, the Free State was urging its agents in Africa to proceed vigorously with the occupation of the Kwango. In July 1889 an expedition led by Frédéric Van de Velde left Matadi for the Kwango, traveled to the capital of "Mwene Putu" at Kasongo Lunda, then retraced its steps and pursued its route eastward to Mwata Kumbana, finally reaching the Lulua at the post of Luebo and returning by way of the Kasai. The following year, Lieutenant Dhanis (who would later achieve greater fame for his part in the "Arab" campaigns of 1892-1895) returned to the Kwango and again visited the Kiamfu of the Bayaka before plunging south below the ninth parallel in a deliberate attempt to secure for the Free State the entire right bank of the Kwango.[46]

In the meantime, Portugal had become thoroughly alarmed and

was hurriedly taking countermeasures of its own. On October 31, 1889, the minister of foreign affairs ordered a mission to be dispatched to the Mwaant Yaav in order to secure his acceptance of a Portuguese protectorate.[47] Portugal also took advantage of the Anti-Slavery Conference which opened in Brussels in November 1889 to offer as evidence of its contribution to the suppression of the slave trade the posts created by Dias de Carvalho during the previous years, but this rather transparent attempt to secure indirect recognition of territorial claims immediately triggered a strong reaction from the Free State as well as from Great Britain, and the entire maneuver petered out.[48] In fact, the winds were now blowing against Portugal and Leopold knew it.

On September 21, 1889, Great Britain had proclaimed a protectorate over the Shire River region where British missions had been active for several years, but which Portugal had always considered a part of her sphere of influence. In October of the same year, Cecil Rhodes's British South Africa Company had been granted a charter which placed no real limits upon its right to extend its authority northward from the Transvaal. Given Cecil Rhodes's unbounded appetite and the political power he wielded in South Africa—he became prime minister of the Cape Colony in July 1890—this meant in effect that Portugal's claims to the interior would now be ruthlessly brushed aside. Indeed, on January 10, 1890, Great Britain sent a strong ultimatum to Lisbon protesting against Portuguese attempts to occupy the Shire valley, and later that year there would be incidents between agents of the chartered company and a Portuguese column.

When Caprivi succeeded Bismarck and, on July 1, 1890, reached a broad settlement of Anglo-German differences over East Africa, Portugal was deprived of whatever nominal support Germany had earlier been willing to grant for her claims over the interior. Thus Leopold could afford to be intransigent with Portugal on the reasonable assumption that he would indirectly benefit from Britain's inflexible opposition to Portugal's transcontinental dreams. For his part, he was at that time on excellent terms with British colonial circles, having concluded in May the so-called McKinnon Treaty with the British East Africa Company. The measure of Leopold's assurance that Portugal could be pushed around without risk was the decree of June 10, 1890, whereby the Free State—without waiting for the return of the Dhanis expedition, which at that time had not even reached the sixth parallel—unilaterally created the District of Kwango Oriental covering all the territory east of the upper Kwango and thus, virtually, the whole Lunda empire.

Yet, if Great Britain had no compunction about twisting Portugal's arm when her interests were at stake and even felt free to appease another major power with an offer of Portuguese territory (as she was to do with Germany in 1898 and later), she had of course no reason to despoil Portugal for the sole benefit of King Leopold, especially since the Free State, once Portugal was out of the way, was bound to be the major obstacle to Cecil Rhodes's northward advance. Already, an agent of the British South Africa Company by the name of Sharpe was on his way to Msiri's court at Bunkeya and, within months, the Free State and the chartered company would be engaged in a headlong race for Katanga.

Under the circumstances, it was hardly surprising that Leopold should be suspicious of British maneuvering: in a coded message of August 21, 1890, the Belgian ambassador to Lisbon indicated that Great Britain had encouraged Portugal to seize Lunda in compensation for the territory she was being forced to surrender to the British South Africa Company.[49] Forty-eight hours after this warning, Portugal raised an official protest, which Brussels of course rejected at once, against the Free State's high-handed decree of June 10. Meanwhile, a long-delayed Portuguese expedition to Lunda under Simão Cândido Sarmento had finally left Malanje, accompanied by fifteen members of the Lunda delegation which the pretender Samadiamba had sent to Luanda at the time of Carvalho's return journey.[50] On September 11 Sarmento encountered the Dhanis expedition at the little village of Mona Samba. Both men, while maintaining cordial personal relations, exchanged formal notes of protest against each other's presence in the area, but Sarmento, who had only eight soldiers against Dhanis's sixty, could not prevent the Belgian officer from continuing his journey to the south or from raising the Free State's flag at Capenda Camulemba.[51]

While violence was avoided in the field, however, some saber-rattling was to be heard in Lisbon as well as in Brussels. The Portuguese foreign minister, Hintze Ribeiro, still confident of Britain's benevolent neutrality, was hinting at a possible use of force to uphold his country's claims over Lunda, while Leopold, not to be out-bluffed, dispatched one of his aides to London with instructions to negotiate in an ostentatious way the purchase of two battleships.[52] It seems unlikely, of course, that either power intended to engage in actual hostilities, and, before the end of the year, Hintze Ribeiro was out of office and the two governments were moving toward a negotiated solution. On December 31 Portugal and the Free State agreed on the agenda of a bilateral conference, which opened in Lisbon on February 19, 1891. The final agreement was reached on May 25.[53] It consecrated the dismemberment of Lunda into two major sections. Portugal secured most of the lands west of the Kasai, including Shinje (Maxinje), Cassassa, Cahungula, and Mataba, but the Free State retained all the area between the Kwango and Kwilu as far south as the eighth parallel — including most of the lands of "Mwene Putu" Kasongo. The Free State also gained control of all the land east of the Kasai, that is, of the Lunda heartland.

The negotiations surrounding the partition of Lunda are a good example of the arbitrary way in which European powers carved up the African continent, but, in the case of Lunda, there was the additional irony of alien conquerors bitterly debating the fate of an empire which had ceased to exist. Much was made in the correspondence between Leopold and his aides of the fact that the 1891 agreement gave the Musuumb to the Free State but, at the time, the capital which earlier reports had described was no more and Mushidi, the new Mwaant Yaav, was living as a virtual refugee on the edges of the Lunda homeland, paying tribute to the Cokwe.

The acquisition of the Lunda heartland had been the indirect result of a conflict which had its real sources much farther west, among the Bena Lulua, in the land of "Mwene Putu," and on the estuary of the Congo. Indeed, the historical core of the Lunda empire had passed into Belgian control even though not a single agent of Leopold had ever set foot in the area, and it would be several years before any attempts were made to occupy it.

3

The Consolidation of Belgian Rule: Initial Patterns of "Native Administration" (1891-1918)

From the very first day that it passed under Leopold's control, it seems to have been the fate of the Lunda heartland to drift into a state of marginality from which it never really re-emerged. The partition of Lunda had been a diplomatic interlude, the codicil to a package deal whose real importance, in the eyes of both parties, had been the determination of the western sections of the Congo-Angola frontier. But, while Free State diplomats were maneuvering Portugal toward a favorable settlement of their claims, Leopold was playing for much higher stakes against a far more formidable opponent. The stakes were Katanga, the adversary was Cecil Rhodes—and behind Cecil Rhodes was Great Britain.

For some time prior to the creation of the British South Africa Company, some British colonial circles had intimated that England should not ratify any demarcation of boundaries in Africa unless they were based on effective occupation. The Lobengula Treaty of 1888 had already given a foretaste of Rhodes's ambitions and methods in this respect. On October 29, 1889, the granting of a royal charter to the British South Africa Company with powers of government over a huge area, whose northern limit was left undefined, revealed the full scope of the threat to the southeastern portion of the Free State. Within a month, Captain Paul Le Marinel had been commissioned to lead an expedition into Katanga (and, if possible, to Lake Nyasa) in order to forestall British penetration. It was to take over two years and three other expeditions (Delcommune, Stairs, Bia) before Leopold's grip over Katanga was assured. In the process, Msiri, the old

Nyamwezi despot who had successfully evaded earlier attempts by Cecil Rhodes's agents to accept British protection, had been shot by a member of the Stairs expedition, and the empire he had carved out for himself had been shattered. More important, in his desperate attempts to secure Katanga, Leopold had granted proprietary rights over the area to a private corporation, the Compagnie du Katanga, in which British interests had shrewdly been permitted to secure a share larger even than that of the Free State. The new corporation (founded on April 15, 1891) financed the two expeditions led by W. E. Stairs and Lucien Bia and took over the Delcommune expedition, which had been funded by the first major Belgian company to operate in the Congo, Albert Thys's Compagnie du Congo pour le Commerce et l'Industrie (C.C.C.I.). In the meantime, three days after the Compagnie du Katanga had been incorporated, the first expedition, under Le Marinel, had finally reached Bunkeya. Before the year was out, Delcommune and Stairs successively reached Msiri's capital and the flag of the Free State was floating over Garenganze.[1]

By contrast with the intense activity and interest displayed over the occupation of Katanga, the news of the settlement with Portugal and the acquisition of the Lunda heartland went almost unnoticed. Much of the excitement over Msiri's empire had of course been generated by the realization that Cecil Rhodes's challenge, unlike that of Portugal, was not to be dismissed lightly, as well as by early reports of Katanga's mineral wealth. Even so, once the threat of British encroachment had been averted, European occupation of Katanga remained so sparse as to be almost nonexistent. In the words of one of the very few officials stationed in the area before the turn of the century, "There never were more than six Free State agents at any one time in Katanga. In 1896, there were three; in 1899-1900, only two. . . . As for missionaries, in 1896 there were [two British Protestant missionaries]. And when it comes to traders, they were so 'elusive' that none could ever be discovered in all of Katanga . . . until 1905."[2]

It hardly comes as a surprise, therefore, to find that no contacts had been established between an agent of the Free State and the Mwaant Yaav before 1896 and that a permanent post was not set up in Lunda until 1903. Of all the intineraries followed to penetrate Katanga, only the northwestern route to Lusambo came within less than a week's march from Lunda, and there was no immediate incentive for Leopold's emissaries to venture into the area. Administratively, of course, Lunda was not even part of Katanga at all: the concession granted to the Compagnie du Katanga did not extend westward to the Kasai River but only to the line of 23°54' east longitude, which follows approximately the eastern limits of the Lunda heartland. Until 1912 when it was transferred to Katanga, the whole pedicle extending southward to Dilolo was administered (to the extent that it was administered at all) as a part of the Congo-Kasai district with headquarters at Lusambo.

But there was of course a more serious reason for this administrative neglect, namely the outbreak of hostilities between the Free State and the Arabs east of Lusambo on the upper Lomami in the autumn of 1892. The conflict, the sequels of which lasted more than a decade, was the product of an ambiguous relationship that had developed during the early years of the Free State and had led at one point to the appointment of the most prominent Arab trader, Tippu Tib, as the official governor at Stanley Falls on behalf of the Free State.[3] Perhaps the most important single source of friction, however, was Leopold's decision (prompted in no small degree by the fast mounting cost of securing effective occupation) to establish a state monopoly over ivory, thereby depriving the Arabs of a major source of income.[4]

The magnitude of the Arab reaction was commensurate with the scope of their political and commercial hold over the eastern Congo, but all private traders were affected — and antagonized — to a greater or lesser degree. Thus, Chief Kalamba of the Bena Lulua, who had developed an enviable trading position for himself and had maintained excellent relations with the Europeans ever since receiving the visit of von Wissmann a decade earlier, rose against the Free State, assisted by the Cokwe, in February 1895.[5] The rebellion was not crushed until the end of June. On June 4 the Tetela and Luba contingents of the Force Publique mutinied at Luluabourg. This first phase of the rebellion ended inconclusively in January 1897, only to be followed less than four weeks later by another mutiny of Tetela and Kusu troops on the march toward the Bahr-el Ghazal in pursuit of one of Leopold's most chimerical dreams.[6] In Kasai and Katanga, the mutinies precipitated a state of general unrest and a protracted series of skirmishes involving a number of mutineers who had scattered into the bush as well as a variety of irregulars and freebooters, most prominent among whom were the ubiquitous Cokwe, who thrived on this sort of situation.

It would no doubt be inaccurate to picture Belgian police operations in Kasai during that period as having been exclusively directed against the Cokwe, but it is nevertheless true that the Cokwe were the most turbulent group in the area and a major nuisance to European officials. Out of twenty-three major "pacification" campaigns occurring in Kasai from 1893 to 1911, ten were directed against the Cokwe, with the Bena Lulua (usually allied with the Cokwe during that period) in second place on the list.[7] Even if we allow for the fact that the term "Cokwe" was being applied rather loosely by Free State officials to all native traders coming from Angola,[8] they seem to have been involved in an unusually large number of clashes with the administration over a wide area ranging from Lake Dilolo to the Kwango region. The state monopoly over ivory, the arbitrary terms of trade which the Free State sought to enforce over legitimate goods, the shoddy quality of commodities available through Belgian companies, and the administration's policy of underpaying the services which they requisitioned, all appear to have contributed to the disgust felt toward "Bula Matari" on the part of all trading communities — particularly the Cokwe and Bena Lulua. An Italian agent in the service of the Free State wrote in 1904:

> Portuguese competition is very strong in the whole Kasai and Lulua region. The Portuguese bring alcohol, salt, textiles, beads, gunpowder, guns and every other kind of merchandise. They pay their carriers one length of cloth per day. Thus, all State goods, which cannot afford such high rates, are at a disadvantage because of this competition. . . . Nor can the Compagnie du Kasai meet this Portuguese competition as the State gunpowder, textiles and other goods are without any value . . . . The Kiokos around here, with their commercial instinct, cannot understand the levying of duties; for years the Portuguese have been trading with them . . . and consequently they are naturally inclined to go over to them since, as they say, the State (M'Bula Matari) is poor and does not pay them.[9]

Actually, most of the clashes involving the Cokwe seem to have occurred in the central and western portions of Kasai and in the Kwango. As mentioned in the previous chapter, this was where the main thrust of the Cokwe trade had been directed since the 1880s. As for the Lunda heartland, it had lost much of its commercial attractiveness with the opening of central Kasai via Kalamba's market and, in fact, "fewer Cokwe had moved east across the Kasai in the 1880s and 1890s than had gone north or south."[10] Nevertheless, when an agent of the Free State first made contact with Mwaant Yaav Mushidi in May 1896, the possibility of exploiting the well-known hostility between Lunda and Cokwe was very much on his mind.[11] The official was Captain Michaux, and the expedition he headed had been organized primarily for the purpose of punishing those chiefs who had sided with the Luluabourg mutineers and to intimidate the rest into an attitude of benevolent neutrality at the least. He had failed to make contact with Kasongo Nyembo, paramount chief of the Katanga BaLuba, had been warmly welcomed by Mutombo Mkulu, chief of the Bena Kalundwe, and was now faced with the diffident attitude of Mushidi, who had never met a white man yet and refused to receive him. Michaux then decided to ignore the chief's reluctance and marched into the capital accompanied by a small escort. Mushidi and his entire court awaited him and, after a short welcoming ceremony, the Belgian officer and the Mwaant Yaav (flanked by his brother Kawele, who was considered the real power behind the throne)[12] retired into the king's compound for a two-hour conference during which Michaux assured the two brothers of his support against the Cokwe. Michaux was extremely impressed with the size and appearance of the fortified capital of Lunda: the distance from the moat to the site of his reception by Mushidi was one kilometer,[13] and he estimated the population of the capital at 30,000, which would have made it the largest agglomeration in the Congo.[14] At the time of this encounter Mushidi was no longer the pathetic refugee who had been forced to flee to the remotest corner of the former empire toward the end of 1887. From the Lunda stronghold of Ine Cibingu, he had begun to extend his authority and had moved his capital to Mwate Itashi, on the right bank of the Kasidishi River, where Michaux found him. Two other Lunda chiefs, Mwene Kapanga and Mwene Mpanda, had also fortified themselves against the Cokwe in their respective fiefs and had already opened successful hostilities against the invaders. Michaux now returned to Lusambo after defeating two Lulua chiefs and was immediately ordered to meet the Tetela mutineers on the upper Lomami with his best troops.

Left to themselves, the Lunda now resumed their efforts against the invaders, but without any decisive results. Mushidi and Kawele tried in vain to reoccupy the Nkalaany and Luisa river valleys,[15] while Mwene Kapanga scored some successes against the Cokwe on the Lulua. The Cokwe regrouped behind Chief Mawoka in the Sandoa area and mounted a counterattack, but they were checked by Mwene Mpanda, and their retreat turned into a major disaster when a flash flood caused a large number of them to lose their guns or drown.[16]

After the limited initial success of this national liberation campaign, a sort of uneasy stalemate appears to have descended on the Lunda heartland. Whether the Lunda offensive had run out of steam or whether, as some writers suggest, the proximity of Free State troops forced an end to overt belligerence is almost impossible to determine. Most sources for this period derive their information from Lunda oral tradition which, for obvious reasons, is inclined to suggest that only the intervention of the Europeans interrupted their victorious advance. This is the version presented by District Commissioner Labrique in his 1922 account of the Lunda-Cokwe conflict,[17] but Edgar Verdick, who in 1903 established the first permanent administrative post in the Lunda heartland at Katola (on the Lulua River at 9° south latitude), found the Mwaant Yaav's capital near the Kasidishi River (i.e., east of the Nkalaany) in a derelict state and noted in his journal: "This is all that remains of the empire of Muata-Yamvo . . . . The Kioko have to a large extent destroyed Lunda centers and, if it had not been for the Free State's intervention, it would not have taken many years before the last man of that race had been taken as slave to the coast."[18] His major reason for selecting the site of Katola for a permanent post was that it was situated in a sort of no man's land between the Cokwe and the Lunda and thus would make it possible to keep the peace between the two groups.[19] Labrique claims, for his part, that after the Lunda war of liberation no Cokwe remained "north of a line running from Sandoa to Kimpuki,"[20] but a report of 1905 refers to the region situated to the south of the caravan route from Kanda-Kanda (in Kanioka country) to Katola as being "occupied by the Kioko race."[21] Indeed, the same Labrique admits, in a letter written a few months after his above-mentioned account of the wars, that "when the Europeans arrived, the Lunda had not reached Sandoa in the reconquest of the lands they had lost."[22] Discounting Lunda oral sources, Edouard Ndua (himself a Lunda) concludes that only the "direct and open intervention of the Congo Free State in this conflict made it possible to restore the Mwaant Yaav's authority over the whole Kapanga region. . . . Prior to 1903, the major portion of the present *territoire* of Kapanga, and the whole region of Sandoa and Dilolo were in the hands of the Tshokwe."[23]

A list of such contradictory statements could be continued, but what

this conflicting evidence most certainly reflects (apart from the superficial nature of official knowledge of the area at that time) is the
almost inextricable imbrication of populations in much of the Lunda
heartland at the beginning of the century. Organized warfare might
have been over, but small-scale operations were still taking place, and
the Lunda communities of the Lulua valley around Katola complained to Verdick that the Cokwe were still guilty of occasional kidnapping.[24] The Cokwe still held most of the territory between Katola
and Lake Dilolo and continued to trade and travel relatively freely
northeast from Dilolo to the lands of Kasongo Nyembo, although the
frequency of military patrols in that latter area was rapidly detracting
from its attractiveness. More important, entire Cokwe communities
had settled throughout the area, either under their own petty chiefs or
with the permission of local Lunda authorities. There is no evidence to
determine whether the latter type of settlement was the result of a
modus vivendi developed after the Lunda war of liberation or merely
the continuation of the sort of agreement under which the Cokwe had
come to live in the Lunda heartland before they had begun to interfere
in the Lunda political process. The latter alternative would seem to be
the more probable since the southern provinces of the Lunda
heartland had not participated very actively in the *reconquista*. It is
reasonable to assume that local chiefs were not particularly distressed
by the disruption of tributary relations, especially since the Mwaant
Yaav was in no position to offer any kind of military protection in
return for tribute and since most of them had had to make individual
arrangements for their own safety. The Willemoes mission to Lake
Dilolo in 1907 claimed that it had encountered only three chiefs who
still acknowledged their allegiance to the Mwaant Yaav.[25]

There was a more serious reason, however, for the faltering of the
Lunda reconquest after its two or three years of initial success. The
counteroffensive had originated from more than one center of resistance, and different operations had been waged by separate forces
under several leaders: the Lunda redoubt of Mushidi and Kawele had
been one such base of operation, but Mwene Kapanga and Mwene
Mpanda had also earned their share of the Lunda victory. Thus,
Mushidi did not emerge from the war of liberation as the undisputed
symbol of national resistance, nor is it likely that he could have
achieved unquestioned legitimacy in any case, given the frequency and
intricacy of succession struggles over the previous forty years.

European intervention unwittingly precipitated the rift between the major protagonists of the Lunda reconquest. When, in April 1903, a group of Lunda bearers requisitioned by a junior Belgian official deserted on the way to the south, Mushidi and his brother Kawele were briefly arrested in retaliation.[26] Upon being released the following day, the two brothers took to the bush and shortly thereafter attacked a small detachment of Free State troops traveling from Dilolo to Lusambo.[27] Thereafter, relations between the Europeans and the Mwaant Yaav were broken. When Verdick traveled to the upper Lulua to set up the post of Katola in May and June 1903, he found that Mushidi and his followers had vacated their capital some time before his arrival and, on his way back to Luluabourg, his expedition was fired upon by a small party of Mushidi's followers.[28] Later that year, the Mwaant Yaav approached the Free State authorities to negotiate a resumption of normal relations with the Europeans,[29] but nothing apparently came of this move for, by September 1904, Kawele had made an alliance with some Cokwe elements and was in open rebellion at the head of some 400 men.[30] He was attacked and defeated by the troops stationed at Katola a few weeks later but managed to escape and remained in the field for five more years.

In the meantime, however, Chief Kapanga had endeared himself to the Europeans[31] and had espoused the cause of a pretender: Muteba, son of Muteba ya Cikomb (1857-1873). Muteba seems to have been something of a nonentity, "decidedly not very brave" (wrote Lieutenant Scarambone) and utterly dependent on Mwene Kapanga, in whose village he gathered his followers to resist Kawele's attacks.[32] The king-maker scored a major political success when Vice-Governor General Costermans, following the advice of Lieutenant Scarambone and District Commissioner Chenot (themselves undoubtedly influenced by Mwene Kapanga's "loyalty") approved the transfer of the administrative post from Katola to Kapanga's village in December 1904.[33] Thereafter, the position of Mushidi and Kawele became increasingly untenable, but Muteba was unable to defeat them, even with the help of Free State troops, and it was not until July 1, 1907, that he was formally recognized by the Europeans as Mwaant Yaav. Even so, Kawele managed to defeat and kill several of Muteba's supporters, and it was only in 1909 that the two brothers were finally captured (treacherously, according to some versions) and killed on Muteba's orders. After a government post had been established at Kapanga, the new

Mwaant Yaav transferred his *Musuumb* to its present site, a few miles east of Kapanga. More than fifty years were to elapse before Mushidi's descendents were able to recapture the throne.

This arbitrary settlement of the Lunda succession did not of course solve the problem of the new Mwaant Yaav's authority, even in the northern portion of the Lunda heartland. As for the southern region, any kind of stabilization, whether in the form of European control or through a resumption of Lunda royal authority, was still largely a matter of wishful thinking. The support initially extended by the Europeans to Mushidi and later transferred to Muteba had been a matter of tactical convenience rather than a policy based on principle. The Free State's concept of native administration had never been very systematic in any case. A decree of October 6, 1891, authorized the recognition of chiefs in areas determined by the governor general but clearly implied that in some areas such recognition might not occur, either because of the intractability of local chiefs or simply because of the disruption of traditional authorities— a condition which undoubtedly prevailed over much of the eastern Congo. Leopold's single-minded preoccupation with the exploitation of immediately available resources, as well as the conditions created by the Arab and Tetela campaigns, had the effect of limiting any sort of indirect rule to an absolute minimum and direct administration was the de facto policy—except of course where, as in Lunda, there was really no regular administration at all.

After the excesses practiced by the agents of the Free State had reached the proportion of an international scandal, Leopold appointed a Commission of Enquiry which, among its other findings, criticized the absence of a coherent native administration policy.[34] The result was a decree of June 3, 1906, which, in its first article, decided rather summarily that all natives of the Congo must belong to a *chefferie* and went on to describe a "chefferie" as consisting of one or several villages placed under the authority of a single chief. Taken literally, the decree of 1906 could mean the fragmentation of traditional authority among a multitude of village chiefs, and, in its desire to provide general guidelines, it adopted as a conceptual yardstick the smallest common denominator of customary government.

When it came to actual implementation in the field, however, the realities of the local situation in Lunda as well as the inevitable distortions of policy occurring along the chain of command combined to produce an adjustable response. Muteba had been formally invested in 1907 and his authority recognized, at least on paper, over an area roughly equivalent to that of modern Kapanga territory; but to the south the situation was still, at best, rather confused. In 1901, in the wake of the Malfeyt expedition, a military post had been set up in the area of Dilolo for the primary purpose of checking the contraband in ivory, rubber, and firearms carried by Cokwe and Mbundu parties.[35] Cokwe trade routes still reached as far north as Kanda-Kanda and there were occasional armed encounters with Europeans.[36] As noted earlier, the Free State could not compete with the business opportunities offered on the Portuguese side of the border and there was, accordingly, a serious concern that the Cokwe would simply remove themselves to Angola if any real attempt was made to bring them under control.[37]

The presence in the area of bands of deserters from the Force Publique (most of them survivors of the Luluabourg mutiny of 1895), whose raidings supplied the Cokwe trade and who were in turn supplied by them with firearms, only made the situation more inextricable. In 1905 the Free State post at Dilolo was practically surrounded by small groups of hostile elements, and communications between Dilolo and Katola were virtually interrupted.[38] The major troublemaker during that period seems to have been a Cokwe chieftain by the name of Kiniama who kept the authorities at bay for at least three years. Lieutenant De Clerck, who took over at Dilolo toward the end of 1905, staged an action against Kiniama in April 1906 but without any success. It took a series of operations between July 1907 and May 1908 against the Cokwe, the Alwena, and the deserters to "pacify" the southern regions.[39] In 1908 the Belgians sealed off the border with Angola, at least tentatively, thereby forcing those Cokwe who remained inside the Congo to bring their goods to local trading posts. During that same year a Portuguese detachment mounted a major operation against the Cokwe of the Moxico area.

The famine of 1910-1912 put a final end to long-distance trading by the Cokwe.[40] Those who remained in the Lunda heartland were a rather heterogeneous group. A substantial number of Cokwe had settled into a relatively sedentary way of life—often with Lunda women—either under their own chieftains or on the lands of Lunda chiefs to whom they paid at least nominal allegiance. The others, found mostly in the vicinity of Dilolo, had been traders or adventurers, usually operating outside, and reluctant to accept any organized authority. For both groups such foci of traditional legitimacy as they might have been willing to recognize lay to the west of the Kasai in Portuguese territory, although scions of chiefly families could be found in the Congo.

None of these intricacies, of course, were fully appreciated by the Belgian colonial authorities at the time, and, as a measure of "law and order" was being brought to the southern region, the problem of a future native administration system for the area still remained to be tackled.

The new decree on native administration passed on May 2, 1910, after the Free State had given way to the Belgian Congo, provided a pattern that could, at least potentially, be applied to Lunda. Article 2 of the new decree provided that recognized chiefdoms could be divided, in accordance with custom, into *sous-chefferies*. Custom would also determine the identity of chiefs and subchiefs, but their authority would be subject to an explicit recognition by the district commissioner, who could also depose them (Article 9). The decree probably came as close to establishing a form of indirect rule as might reasonably be expected in a territory that had just lived through a full generation of disruptive crises from which no traditional society had emerged intact, but the underlying philosophy was clearly that chiefs were auxiliaries of the colonial administration rather than autonomous authorities: not only could they be removed for reasons of incompetence, but they also received a salary (to which traditional tributes could be added) and were explicitly enjoined to take orders and directives from the European authorities (Articles 9, 13, 15, 16).

As it was, the decree of 1910 could easily be made to fit the traditional structure of the Lunda empire. There were only two problems as far as the Belgians were concerned, but they were crucial to all future policy developments: could the Lunda empire be restored? and if so, should it? Having been instrumental in Muteba's accession to the throne, the Belgians tended to assume, not altogether incorrectly, that whatever power he might still be able to wield would be derived in large part from the ostensible backing of the colonial administration. In doing so, however, they overlooked the extent to which the office of Mwaant Yaav might still retain some of the institutional prestige not extended to the person of Muteba himself, especially after the shattering experience of Cokwe hegemony. For similar reasons, they also tended to expect from the Mwaant Yaav the sort of dynamic, activist leadership which he was unable to provide, either on the basis of tradition or in the new context of the colonial situation. The amount of resistance offered by the Lunda to the establishment of Belgian rule had been minimal, to be sure, and there was accordingly no major political reason why the colonial authorities should have wished to dismember the empire, but, in the event, the empire had already been dismantled, and the issue was really whether they should forcibly intervene to ensure its revival. The tenuous character of the Mwaant Yaav's claims to authority over non-Lunda elements was perceived from the start. In 1913 District Commissioner Gosme wrote:

> In the region of Dilolo, many reputedly Tshiokwe villages have a wholly Lunda population and only the chief and his family are Tshiokwe. Some of these villages recognize the Mwata-Yamvo's authority. . . . But this allegiance should not fool anyone; it is purely superficial and would vanish rapidly if the Mwata-Yamvo ever dared to attempt any act of real authority. It derives primarily from the fact that the Mwata-Yamvo, who has been officially invested, is protected by the colonial authorities who are interested in maintaining his authority. We are the ones they really fear behind the Mwata-Yamvo.[41]

Another drawback — which, under a different policy, might actually have been an asset — regarding the potential resuscitation of the Lunda empire was its size, even without its former eastern and western provinces. A reunified Lunda heartland would have covered several territoires and might even have straddled provincial boundaries. But Belgian field administration (like the French) was organized in such a way as to make any departure from its own institutional logic almost unthinkable. When in early 1916 the governor of Katanga wrote to the district commissioner for Lulua that it would be acceptable, as an interim measure, to have a chefferie extending over more than one territoire supervised by the official within whose jurisdiction was found the traditional center of the chefferie, the governor general stiffly reminded him that such a provision was "inadmissible," since "the area of jurisdiction of District Commissioners and Territorial Administrators can only extend over the territorial unit which they are specifically responsible for administering."[42] In other words, a reconstructed Lunda state would have to be under the concurrent supervision of several officials of equal rank.

In 1910 the Comité Spécial du Katanga (C.S.K.), a semipublic agency which had taken over the joint administration of public and private holdings in the conceded area, turned over its governmental prerogatives to the colonial government, and Katanga was turned into a province of the Belgian Congo under a vice-governor general, although that latter position continued to be filled, ex officio, by the C.S.K.'s senior field representative until 1933. On March 28, 1912, the territory extending from the western limit of the Katanga concession (23°54' east longitude) to the Kasai River boundary with Angola was transferred from the province of Congo-Kasai to Katanga.[43] The Lunda heartland was now welded to a province whose fate it would so decisively affect some fifty years later.

The transfer of Lunda to Katanga took place at a most inauspicious time. Plantation rubber had appeared on the world market. Asian plantations had exported 8,200 tons in 1910, 28,500 in 1912, and 47,500 in 1913. The bottom fell out of the market for native rubber, which had long since replaced ivory as the Congo's chief natural resource. In southeastern Katanga, copper production, which started in 1912, soon filled the gap left by the collapse of rubber, but no such alternatives were available in Lunda, where rubber had been virtually the only source of monetary income available to the Africans. The Compagnie du Kasai closed down five of its nine trading posts in the area in 1913.[44] There was a new outbreak of hostilities between Lunda and Cokwe, caused in part by rumors of a possible restoration of the Mwaant Yaav's authority over the entire region, and generalized unrest spread through the area around Dilolo. The situation was not improved by the unscrupulous and abusive behavior of Belgian agents at Dilolo, which caused them to be brought to trial. Contraband was rampant in the Dilolo area and taxes had reportedly never been paid.[45] Native administration, it was felt, would have to start from scratch:

> We are only just beginning to penetrate this area peacefully. Until now, except for Kapanga and the immediate vicinity of government posts, it is fair to say that there had been no occupation, peaceful or otherwise. Natives ignore government posts if they are any distance from them. Everything remains to be done in this respect.[46]

The only hopeful factor, District Commissioner Gosme went on, was the presence of the Mwaant Yaav, who "is loyal and as helpful as he can. . . . He may be said to reign over nearly the whole area comprising the territories of Kapanga, Dilolo and Kafakumba. . . . I prescribed," he added somewhat hopefully, "looking into the possibility of having him recognized as paramount chief of the whole area, without hurting the natives' feelings on this score."[47]

Yet, even among the Lunda, the Mwaant Yaav's authority was still far from being generally accepted, particularly by surviving members of the deposed royal line. Tshipao, a half-brother and wartime companion of Mushidi and Kawele, persistently denied Muteba's legitimacy and in 1912, hoping perhaps to advance his cause by moving closer to the seat of European power, he settled with his followers near the newly-created district headquarters of Kafakumba, claiming for himself the title of "Mwaant Yaav a Rul" (Mwaant Yaav of the East). District Commissioner Gosme had him locked up at Dilolo until he agreed to surrender his claims to the throne, in return for which he was left in charge of an extensive fief around Kafakumba and permitted to retain his spurious title. In June 1913, Gosme brought Muteba to the district headquarters to receive Tshipao's homage and used the occasion to show him ostensible deference, thus hoping, presumably, to advance his stated objective of "establishing a definite basis for the reorganization of the whole Lunda race into one vast chefferie."[48] Similar ceremonies were staged in other parts of the district.

Another, more serious obstacle in the district commissioner's path was the extensive imbrication of Lunda and Cokwe communities, as a result of the circumstances of previous decades. "Batshioko, Lunda and even Baluba are totally jumbled," Gosme acknowledged, "and it will be very difficult to organize them into separate chefferies."[49] What was even more disturbing, his successor later added, "the authority of those notables susceptible of being selected as chiefs is rejected by many inhabitants," since any small groups wishing to do so "can only too easily break away from the main group simply by moving away from known trailways."[50] This was particularly true of the Cokwe, and one can appreciate the sense of frustration which led Gosme to decree that "the Tshiokwe shall of course have to return, even against their will, to the chefferie to which they belong." He added that they would have to be persuaded that "we have no intention of subjugating them to a chief of our own choice but only to reconstitute the major original divisions of their race." But then, as Gosme candidly admitted, "as long as [the Cokwe] remain divided in small independent hamlets . . . we will have a hard time tackling them. They like this anarchy which permits them to give way with impunity to their fierce instincts and to their taste for piracy."[51]

Under the circumstances, it could hardly be expected that there would be rapid progress toward the establishment of a coherent native administration system, and the outbreak of the war in Europe in 1914 was hardly conducive to the pursuit of long-term objectives. The Cokwe problem was at a standstill, and the only visible change in the situation was the short-lived use of the term, "Territoire des Tshiokwe," to designate the southernmost portion of the region; but, as official reports admit somewhat ruefully, there was no real contact between the population and the junior agent then manning the post at Dilolo.[52] The amount of money circulating in the district had been drastically curtailed as a result of the rubber crisis, and Portuguese traders, who were virtually alone in the area, increasingly resorted to some form of barter. Tax collection fell from 46,529 francs in 1913 to 33,922 francs in 1914, a drop of 27 percent.

A new attempt to organize the Cokwe was made in 1915, this time by herding them together more or less arbitrarily under the authority of Lunda chiefs, who in turn were subordinated to the Mwaant Yaav. One can understand the slowly mounting exasperation—which can be followed in the official reports—of District Commissioner Gosme with the recalcitrant Cokwe. Other circumstances lent themselves to this rather sweeping experiment. A circular issued by the governor general on July 24, 1915, reminded all officials that chiefs had a dual character: as parts of the colonial administrative hierarchy, they were entitled to a salary, but as traditional rulers they could be entitled to customary tribute.[53] Through the bias of tribute, the authority of the Mwaant Yaav could, at least in theory, be restored over vast portions of Lunda, irrespective of the imbrication of Lunda and non-Lunda elements throughout most of the region. Several officials had already noted the willingness of some Cokwe communities, especially on the upper Lulua, to acknowledge the nominal paramountcy of the Mwaant Yaav, at least verbally. The concept of *sous-chefferies*, introduced by the 1910 decree on native administration, could be utilized to install lieutenants of the Mwaant Yaav throughout the area. "Faced with the systematic refusal on the part of Cokwe communities to recognize official chiefs," the district commissioner wrote in 1915, "I have decided to establish the authority of the Mwata-Yamvo over most of these groups, whether they like it or not."[54] And so it was. After the district headquarters had been transferred to Sandoa, the Mwaant Yaav was invited to tour the region in the company of the district commissioner and to nominate a resident chief for the area east of the Lulua. He selected a member of his own clan, Kabokono, who was installed as Mwene Matanga and put in charge of a sous-chefferie in April 1916. Tshipao, the former pretender, had of course already been installed earlier in a sous-chefferie of his own on the eastern edge of the Lunda heartland as political compensation for renouncing his claims to the Lunda throne. Successively, following the same pattern, the Mwaant Yaav was invited to designate subchiefs for the areas to the north and south of Sandoa — Muteba, Tshibamba, Mbako, and Muyeye — and again he appointed close relatives to these positions. We have no contemporary statistics regarding the respective numbers of Lunda and Cokwe living in these areas, but figures compiled in 1929 indicate that at that time the number of Cokwe equaled or exceeded that of Lunda in all but one of these four sous-chefferies.[55]

The new policy clearly had the full approval of the Katanga provincial authorities. In a letter of January 21, 1916, addressed to the district commissioner for Lulua, the vice-governor general for Katanga specified that "in order for [the Cokwe effectively to be subordinated to the Mwata-Yamvo] you will have to be careful to determine the tribute that will have to be paid," and suggested that such tribute should "consist of local produce . . . which the Mwata Yamvo might sell to traders at Sandoa." In subsequent correspondence, the vice-governor's position was explained as follows:

> The most simple and practical solution was to determine or to approve, at the time of [the subchiefs'] appointment, the nature of the tribute to be paid by the Tshiokwe to their new suzerain. Without the tribute, without an obligation, the subjugation of the Tshiokwe to the Mwata Yamvo would have been purely nominal and we would have reaped no benefit from the new organization of the land.[56]

The view from Léopoldville was somewhat different, however. In his letter of April 1, 1916 to the vice-governor for Katanga, the governor general was of the opinion that

> it is undesirable that we should get involved in the collection of tribute demanded of the natives by their chiefs and sub-chiefs. . . . You are too much aware of the requirements of native policy for me

to have to draw your attention to all the excesses in which chiefs will indulge under the excuse of tribute.[57]

Nor was District Commissioner Gosme, who had done so much to build up Muteba's position, quite as sanguine as he had once been regarding the Mwaant Yaav's performance. In August 1916, he admitted that

> the Mwata Yamvo relies too much on his emissaries who take advantage of the mission entrusted to them to exploit the population . . . . This chief has been spoiled by the Europeans who have not been sufficiently distant with him over the past year or so; too much has been given to him and too little work has been asked of him.[58]

The Belgian *administrateur* for Kapanga, Hupperts, confirmed that "the Mwata Yamvo's *capitas* (emissaries) terrorize the villagers" and even tried to suppress them for a while, but eventually had to settle for a reduction of their number to thirty, which he hoped would permit a stricter control of their operations.[59]

To the Lunda aristocracy, however, the new policy was a providential and somewhat unexpected boon after the years of uncertainty. The ritual investiture of Muteba in the Nkalaany Valley in the fall of 1916, nine years after he had been recognized as Mwaant Yaav by the Belgians, probably reflected the Lunda establishment's new-found sense of confidence as well as the recognition of the political need to reinforce the paramount's position in the eyes of the administration — which indeed acknowledged the ritual consolidation of Muteba's authority by awarding him the bronze medal reserved for senior chiefs.[60] To the Cokwe, however, the new policy was an unacceptable attempt by the colonial administration to alter in favor of the Lunda the balance of forces which had evolved in the heartland of the former empire as a result of the Lunda-Cokwe wars. To the extent that the area affected by the new policy was largely restricted to the region around Sandoa and, more important perhaps, to the extent that most Cokwe had been accustomed for years not to pay any great attention to chiefs of any kind, the new policy's impact was somewhat delayed. But many Cokwe notables took it quite seriously, such as, for example, the future Chief Tshisenge, who simply removed himself and his followers to Angola when Belgian officials escorted Muteba on the western, Cokwe-held bank of the Lulua river.[61]

The new Lunda chiefs for their part left no doubt that the Cokwe

living under their jurisdiction were being reduced to the rank of subjects and that the collection of tribute should be interpreted as a visible
sign of political vassalage. It was of course refused on the same gounds,
and a Belgian official, writing in retrospect of that period, could
claim, perhaps too sweepingly, that the policy introduced in 1915 had
been "the original cause of the political anarchy in the whole Tshiokwe
group."[62]

Only the Cokwe had settled in the midst of the Lunda heartland,
and they alone had shown such reluctance to let themselves be organized into native administration units; but other groups located on the
periphery of the heartland and who had in the past fallen within the
Lunda orbit were also affected by the new policy although to a lesser
degree. The Lwena, the northernmost members of the Lovale family
which is mainly found in Zambia, occupy a narrow strip in the southwestern corner of modern Katanga. These people move seasonally to
fish near the headwaters of the Zambezi. Thus, economically as well as
culturally, their traditional orientation had been to the south, and
they tended to regard their residence in the Congo as marginal. Not
surprisingly, therefore, when the district commissioner escorted the
Mwaant Yaav on a tour of the Dilolo area—a procedure already
utilized on previous occasions—the head of the senior Lwena clan in
the area, Katende, interpreted the trip (quite correctly) as a prelude to
his subordination to the Lunda paramount and moved his village
across the border into Angola, where he remained until the Belgian
administration in 1932 finally recognized the autonomy of his people
from Lunda overrule.[63]

To the north, emissaries of the Mwaant Yaav attempted to collect
tribute from southern Kete villages of Luisa territory, which indeed
had once been linked by a semitributary relationship to the empire but
had not been transferred to Katanga in 1912 and consequently were
now under a new provincial administration, one which took a
curiously proprietary view of "its" villagers and was not prepared to
tolerate the intrusion into Kasai of the representatives of a "Katanga
chief." The Kete issue, which had initially involved only the administrateurs de territoire of Kapanga and Luisa, eventually escalated,
after some acrimonious correspondence, to the level of district and
even provincial officials. It was finally settled between 1924 and 1926
when Muteba's successor, Kaumba, was forced to surrender his claims

over the southern BaKete, but incidents continued for some time thereafter.[64]

To the southeast the southern Lunda (or Ndembu) had been organized under Kazembe Mutanda and Musokantanda, but, like the Lwena, their major concentration was to the south of the border, and Musokantanda, for one, preferred to reside in Northern Rhodesia. In any case, both areas had been independent from the Lunda empire since the eighteenth century, and, although Kazembe Mutanda recognized historical links with the Mwaant Yaav, neither he nor, a fortiori, the Musokantanda line, which had been founded by a Lunda adventurer, had been tributaries of the emperor at the time of the Cokwe invasions.

Even within the heartland, as subsequent developments were to show, the loyalty of the Lunda *yikeezy* selected by the Mwaant Yaav was not as complete as might have been expected. Many of them had been accepted only grudgingly by the local population[65] and found themselves in an uncomfortable position. Their initial selection by Muteba clearly marked them as lieutenants of the Mwaant Yaav, but those aspects of their expected performance which were regarded as primordial by the Belgian administration had little or no equivalent in the traditional process. To the extent that they were preoccupied with the perpetuation and possible institutionalization of their profitable new positions, excessive dependence on an ageing sovereign could be a double-edged sword. Indeed, many local headmen who had themselves aspired to be selected as chiefs had not entirely given up the hope of seeing some of these appointments reversed by the Mwaant Yaav. This factor could, and did, give the Lunda paramount a measure of leverage over the new chiefs, but could also induce these same chiefs to seek direct support from the colonial administration. In fact, many of them soon came to recognize the fact that their newly acquired power, as well as that of the Mwaant Yaav, ultimately depended upon the Belgian authorities, and that the paramount's disfavor might not be sufficient to dislodge a *sous-chef* who met the demands of the colonial officials, or vice versa.

The heartland's subdivision into several territoires gave the chiefs direct access to Belgian officials. While an administrateur de territoire was responsible for supervising the Mwaant Yaav's own domain, as defined in the 1907 warrant, the new or emerging chefferies en-

trusted to Mwene Matanga, Muteba, Mbako, Muyeye and others were
under the direct authority of other administrators of equal rank posted
at Sandoa, Dilolo, or Kafakumba. The fact that the district commissioner himself was based at Sandoa (after having been stationed at
Kafakumba and, earlier still, at Dilolo) rather than at Kapanga could
also work to the advantage of local chiefs. At the same time there were
recurrent complaints on the part of successive colonial officials (as
late, in fact, as 1950) to the effect that the Mwaant Yaav's own chefferie was "absolutely too large to be usefully directed by a single man"
and should be divided into several sous-chefferies.[66]

Actually, none of these potential problems had developed into a
major issue by the end of World War I, but the tide which had run for
several years in the Mwaant Yaav's direction was now beginning to
turn perceptibly. By the middle of 1917, Belgian officials had come to
realize that many sections of the Lunda heartland could be organized
only under non-Lunda chiefs. In September, Labrique, who had
succeeded Gosme as district commissioner, asked the administrator for
Kapanga to sound Muteba's feelings regarding the possible organization of some of the Cokwe communities under their own chiefs.

> I want you to insist upon the importance I attach to (this problem),
> for I assume that such a project should not please the Mwata Yamvo
> too much, since he has consistently turned his lands over to his fa-
> vorite *capitas*. This manner of proceeding offers no disadvantages
> when Lunda populations or small heterogeneous groups . . . are in-
> volved . . . . [But] when dealing with the Tshiokwe, it is impossible
> to ever achieve any result by applying the Mwata Yamvo's system:
> those people will never trust a chief who is thrust upon them if he is
> not of their own race. . . . The number of people escaping across the
> Kasai river shows the results of this policy. Furthermore, the Tshiok-
> we on the Lubilash are now *asking* for a chief. . . . As this whole
> area belongs to the Mwata Yamvo, he is entitled to parcel it out as
> he sees fit (but) no Lunda live in this portion of the Territory . . . .
> Such a chefferie would be ruled by one of Tshisenge's descendants,
> to be designated by the Tshiokwe. Another one of Tshisenge's de-
> scendants came to see me lately; he claims to be chief of the Tshiok-
> we to the South of Sandoa and I authorized him to settle with his
> people.
>   It is of course understood that such chefferies would remain trib-
> utaries of the Mwata Yamvo and that the Tshiokwe who are includ-
> ed in the jurisdictions of Mwene Matanga or Muyeye would under

no circumstances be allowed to emigrate from their present residence . . . . Do not rush things, but make him understand quietly that this matter seems so important to me that I would be prepared to over-ride him.[67]

The district commissioner's ideas were not put into practice until some five years later (by which time Labrique himself had come to change his views almost completely), but the obvious limitations of the earlier policy with respect to non-Lunda groups were clearly beginning to be perceived. At the same time, to the very extent that this policy had apparently succeeded in producing some reasonably effective Lunda chiefs, it was bound to diminish to some degree the Mwaant Yaav's importance in the eyes of the Belgian administration and to lead to his being bypassed by colonial officials in their dealings with local chiefs, particularly since Muteba was now physically and mentally declining.

Meanwhile, in Belgium, the end of World War I brought to power a new minister for colonies. Louis Franck, a Flemish member of the Belgian Liberal party has been described as "Belgium's Lord Lugard"[68]—a label which is not entirely inappropriate and which he would undoubtedly have been proud to wear. Franck is mostly remembered for his circular of November 8, 1920, on "current and future policy regarding the organization of native populations," which he prefaced as follows:

> To disorganize native life in order to try to shape it after our European concepts is to create anarchy. We must continue . . . to make increasingly good use of native institutions, to respect and to develop them in all fields, even if they contradict our ideas, our concepts and our mores, so long as they do not contradict essential principles of humanity and justice. Native populations are the basic wealth of a tropical colony. It is a fallacy to believe that they can advance on any base other than that of their own customs and organization.[69]

The words could have been those of Lugard, but the actual policy was much more inclined toward interventionism. Its major innovation was the creation of *secteurs*, an artificial native administration unit bringing together under a single chief populations who lived either in segmentary societies or in communities too small to be autonomous and who belonged—preferably, but not necessarily—to a single ethnic group. These units were by definition nontraditional and their chiefs, while they might have customary authority over one of the component groups, were of necessity, in the words of Franck, "the auxiliaries and representatives of our administration."[70] The practical need for secteurs was not questionable; there were 6,095 chefferies in 1917, many of them obviously too small, and the multiplication of sectors (especially after 1933 when the system was generalized) made it possible to reduce their number to 1,070 in 1940 and to 476 in 1950,[71] but it would seem excessive to speak of indirect rule to describe this policy. In those areas where large-scale traditional states (*grandes chefferies*) could be found, on the other hand, Franck did recommend that chiefly powers be maintained, even reinforced, and that the breakup of such large units be prevented. Like other disciples of Lugard after him, however, Franck felt that indirect rule could be applied not only to traditional polities that still represented living realities (as had been the case with the emirates of northern Nigeria) but also in areas where such polities could still be "revived," as he put it.[72]

But, apart from his admiration for Lugard, Franck was also (at a time when Belgium was coming to grips for the first time with its ethnic problem) an advocate of Flemish cultural autonomy as well as a statesman of the era that produced the Treaty of Versailles — and also, more to the point, the Treaty of Saint-Germain which registered the breakup of the Hapsburg Empire. To Franck and to many of his contemporaries the autonomy of "national" groups was a hallowed principle, and the form of indirect rule which he advocated was intended to reconcile such "legitimate" aspirations with the necessities of colonial administration. In this perspective, the multinational Lunda state was bound to be viewed in the same unfavorable perspective as the former Austro-Hungarian empire and Franck's directives could easily be interpreted as an encouragement to free the "subject races."

The full effect of Franck's policy, and the full weight of its ambiguities, were not felt in Katanga until the early 1920s. For some time prior to that period, however, Belgian authorities in Katanga had begun, as seen above, to show some reservations about the indiscriminate extension of the Mwaant Yaav's authority over non-Lunda areas. In a memorandum written in November 1919, a senior provincial official noted that, while the administration should reinforce the power of the chiefs over their subjects through its intervention and support, it should avoid at all cost "creating artificial links of subordination" which could only have unfortunate effects in the future.[73]

The lack of official support for the Lunda pretensions in the above-

mentioned incidents about the southern BaKete did not become evident until the mid-twenties, but Mwaant Yaav Muteba's attempt to
intervene in the succession at Kayembe Mukulu a few months before
his own death led to a rather ignominious put-down for the old paramount on the part of the administration. Kayembe Mukulu, like the
barbarian states on the Roman *limes,* had enjoyed a semiautonomous
status as an eastern march of the old empire and, over the years, its
largely Luba population had been gradually swamped by Cokwe infil-
trations, leaving the true Lunda a distinct minority, far behind the
other two groups. Chief Kayembe Mukulu, second of the Kalenda line,
died in early 1920, and Muteba backed a descendant of the previous
dynasty, Ndaye Fwana Tshikweji, against Bombolongo, brother of the
deceased chief. Local dignitaries, however, supported Bombolongo
against the Mwaant Yaav's choice, at which point the administration
intervened to warn the Mwaant Yaav in no uncertain terms to keep out
of the succession dispute.[74]

Muteba died on August 31, 1920, before the full dimensions of the
administration's gradual reversal toward the Lunda state had really
become manifest. The monarchy's prestige and the power of the
Lunda aristocracy were incomparably stronger upon his death than
they had been when he had been propelled to the throne with the help
of Mwene Kapanga's intrigues, but these gains had been due to the
deliberate support of the colonial administration, prompted in turn by
their desire to "organize" quickly and effectively the populations of
southwestern Katanga, much more than to Muteba's own maneuvering skills. Faced with the administration's gradual reconsideration
of its earlier policies, Muteba's successor would find that it took all the
limited leverage which an African chief could wield in a colonial system merely to retain some of the privileges which had been thrust so
liberally upon the previous Mwaant Yaav. As a matter of fact, the very
conditions under which Muteba's succession was decided gave the
administration a ready-made opportunity to trim the future Mwaant
Yaav's prerogatives, thus paving the way for the policy of "national
autonomy" which was to be formally articulated three years later.

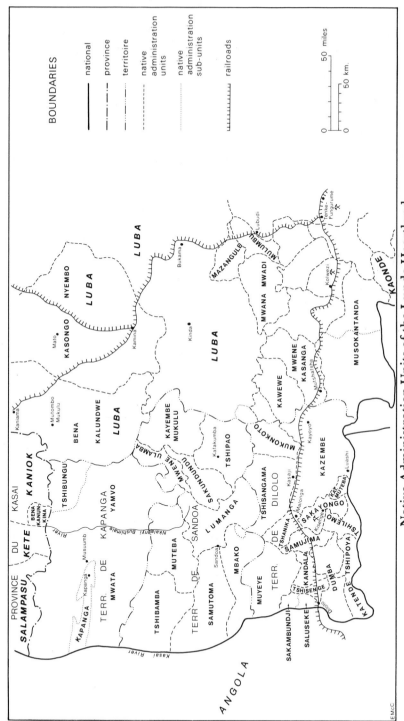

Native Administration Units of the Lunda Heartland
During the Colonial Period

4

Prelude to Dismemberment:
Administrative Rationality and the Policy
of "Ethnic Autonomy"
(1919 - 1932)

Colonial rule has been called a "working misunderstanding," but what made it work in most cases was the lack of alternatives on the part of the colonized and, by contrast, the relative ease with which superior material power made it possible for the colonial ruler to devise alternative policies or merely to reverse himself with virtual impunity—at least in the short run. The colonizer could afford, to a surprisingly large extent, to be inconsistent in matters of native administration, while traditional elites found it virtually impossible to follow a long-term strategy toward a power structure which seldom regarded the administration of African communities as an end in itself and which could resolve its contradictions through the *ultima ratio* of force.

Mushidi's deposition and the accession of Muteba to the throne had started as an internal Lunda problem, a chapter in the unfolding series of court intrigues which had chronically afflicted the empire since the death of Naweej II, but its real significance—whether or not it was realized by Mwene Kapanga, the "kingmaker"—lay in the fact that the issue had been settled by the Europeans. To them it mattered relatively little in the long run whether the throne was held by the son of Mbumba or by that of Muteba ya Cikombe; the important fact was that neither could reign without their implicit support. In that perspective, Muteba's accession had served the colonial power; now, his death could be politically useful.

Two claimants competed for the succession. One was old Tshipao, the son of Mbumba who had claimed the throne against Muteba in 1912 and who had been styling himself the "Eastern Mwaant Yaav"; the other, Kaumba, was forty years old upon the death of Muteba, who had made him his *Nswaan Mulopw*.[1] Kaumba was selected to succeed Muteba by court dignitaries in September 1920, but their decision was immediately challenged by Tshipao. His claims were backed by some of the Lunda subchiefs who had been delegated by the late Muteba to administer the southern regions, notably Mwene Matanga and Muyeye. The matter was submitted to District Commissioner Labrique, who in turn interviewed Tshipao. In the process of arguing his case, the old pretender suggested that, since the Europeans would not tolerate a war of succession, each local chief should be permitted to pay tribute to the candidate of his choice, to which the district commissioner reacted by observing that this would be tantamount to recognizing the principle of self-determination within the empire. "I told him," Labrique writes, "that this would be admitting the free determination of land by its occupants, and that the Cokwe, Luena, etc. might no longer pay tribute to the Lunda. To this he replied in the affirmative."[2]

Having assured Tshipao that he would investigate the matter, the district commissioner asked his subordinate at Kapanga to prepare a background study of Lunda succession procedure but outlined the political dimensions of the problem in the eyes of the administration by adding:

> Kaumba is more interesting from the European viewpoint . . . . But it might be possible to reconcile [the natives'] rights, customary law and our interests. The Lunda chefferie is too large — heterogeneous — thus lacking in cohesiveness and if what Tshipao claims is accurate — that the populations of the Lunda empire have the right freely to determine their fate — this might be a unique opportunity to partition the chefferie.[3]

Reviewing the alternatives for the vice-governor of Katanga on the same day, the district commissioner wrote:

> Either Kaumba will be elected or Tshipao might be. But in either case, there will be dissidents who will refuse to recognize the one or the other. I think it would be bad policy for us to force the one who will be elected on the entire Lunda, Luena, Batshiokwe etc. area. This empire is too extensive to be administered competently. Most of the southern chiefs, Mwene Matanga, Muyeye, Mbako, and Tshipao himself . . . would be favorable to a partition. If the northern group shared this opinion, I think it would be the most elegant solution.

Tshipao claims that anyone is free to choose his chief, no matter what race he belongs to. This assertion leaves me doubtful . . . but if Tshipao admits this principle, we could as of now give their freedom to important groups, like that of Tshisenge's Batshiokwe at Kafakumba, and slowly prepare the future creation of chefferies as homogeneous as possible in anticipation of Tshipao's death.[4]

The reaction from provincial headquarters was quite explicit:

I approve the final considerations of your letter . . . except that there is no need to wait for Tshipao's death to release from servitude the populations which were imprudently subordinated to the Mwata-Yamvo a while ago. . . . I still think that the Mwata-Yamvo's rights extend only over a portion of the Balunda; and I certainly could recognize none other than those except on the basis of satisfactory evidence derived from a report of enquiry. It is true that the situation in the district is quite complicated, due to the widespread imbrication of populations [tribes], but this is no reason for us to leave the solution of the problem to a native chief; this would amount to an abdication of European authority . . . . The Vice-Governor General stressed the fact that we must resort to patience because the proud Batshiokwe tribe, convinced rightly or wrongly of its superiority over the Balunda, never made a sincere submission . . . . After all, occupation of this region under the Free State found the Balunda in the northern part of the district. By demanding the benefit of the situation which existed at that time, the Batshiokwe advance territorial rights which we must recognize. The sous-chefferies to which you refer were all created on the principle that the Mwata-Yamvo has claims over the whole country, whereas this has never been proved. It may be, Mr. District Commissioner, that, as you say, "this policy will create a trend exactly opposite to what existed previously," but injustice and oppression cannot be condoned by the authorities; the path that was followed must be abandoned as soon as possible. I don't believe we have to fear that populations which accepted alien chiefs through fear will be hostile to a recognition of their own headmen.[5]

It soon became clear that Tshipao's claims, no matter how justified they might have been, were not going to prevail in the eyes of the traditional electors, but the administration was in no particular hurry to recognize Kaumba. In a letter dated February 3, 1921, the vice-governor general instructed the district commissioner to withhold formal recognition of the new Mwaant Yaav until a decision had been reached regarding the overhaul of the native administration structure of Lunda.[6] Kaumba, for his part, was understandably anxious to establish his position as soon as possible. To this effect, he traveled to the district headquarters, offered his full cooperation to the district commissioner to help him untangle the complex land issues raised by the new policy, indicated a desire to learn French, and even suggested that he might move his residence to a place near Sandoa.[7] The district commissioner was visibly impressed with Kaumba and, without openly contradicting his superiors, increasingly espoused the Mwaant Yaav's viewpoints and claims.

Apart from his relations with the administration, however, Kaumba still had to secure acceptance by the Lunda aristocracy and to isolate Tshipao's backers. Mwene Matanga, who had been appointed by Kaumba's predecessor to be his representative in the Sandoa region and had been the first Lunda subchief recognized under the 1915 policy, had been favorable to Tshipao and was, by virtue of his location, in a particularly good position to have access to the district commissioner. During his visit to Sandoa, Kaumba pointedly inquired of the district commissioner whether he would be allowed to reverse the appointments made by his predecessor. With the backing of the Belgian official, who had come to feel that the administration was "on the wrong track" and that only a recognition of Lunda custom, "however different it might be from the rest of the Congo,"[8] could ensure a stable native administration program, the new Mwaant Yaav also lent a sympathetic ear to the claims directed by local land chiefs against some of the resident *yikeezy* selected by Muteba. Thanks to his intervention, one of them, Lumanga, was permitted to pay tribute directly to the Musuumb instead of through Mwene Matanga's intermediary. Another *mwaantaangaand,* Kamulemba, openly announced that Mwene Matanga would soon be deposed, whereupon the latter apparently lost his temper, insulted Kaumba, and was promptly fined three months' salary by the district commissioner. Thereafter, Mwene Matanga's position with the administration became increasingly shaky, and in 1924 he was deposed in favor of Lumanga.[9]

A similar scenario was enacted some three months later when the Mwaant Yaav and some of his court dignitaries were invited to pronounce on land rights in the Dilolo region. The only recognized Lunda chief in the area, Muyeye, had also reportedly favored Tshipao, but now, since Kaumba's accession, a number of land chiefs were openly

flouting his authority and demanded the right to pay direct tribute to
the Lunda court. The local branch of the chiefly family of
Sakambundji, which claimed the traditional dignity of *Sanama* (once
held by Mbumba), was given the same prerogative, and its lands were
eventually organized as a separate sous-chefferie in 1925, despite the
fact that they consisted of a mere handful of villages on the Belgian
side of the Kasai River.[10] The same pattern prevailed in the southern
portions of Tshipao's domains, where several direct tributaries were
recognized, two of whom (Mukonkoto and Tshisangama) were also
made subchiefs five years later. Farther to the south, Kaumba
endeavored to use his influence on behalf of Chief Musokantanda
Mwanza of the Ndembu, who had removed himself to Angola several
years earlier and in his absence had been replaced by the Belgians with
a chief named Kalyata who refused to accept the Mwaant Yaav's
suzerainty.[11] Kaumba also did his best to discredit in the eyes of the
colonial authorities a Cokwe headman named Tshisenge Kapuma who
had been attempting for some years to rally his fellow tribesmen settled
in the east central portion of Lunda in order to gain official recognition for an autonomous Cokwe unit.[12]

In the north, Kaumba managed to enlist the district commissioner's
support behind his claims over the southern BaKete villages (see Chapter 3) and over the lands of the Ine Cibingu (Tshibungu)—Mushidi's
redoubt during the Cokwe invasions—which the vagaries of colonial
administration had included in the Lomami district but which, unlike
the Kete villages, were eventually transferred to the Lulua district and
placed under the Mwaant Yaav's paramountcy in 1923.[13]

Not all Belgian officials in the district were equally favorable to
Kaumba, however. Like many field administrators before and since,
the young official stationed at Kafakumba tended to back "his" chiefs;
he had a high opinion of old Tshipao and felt that the colonial
authorities "could not go on maintaining the former Lunda empire"
over Tshipao, Kayembe Mukulu, and the east central Cokwe (all of
whom were under his jurisdiction). "Must we continue," he wrote in
mid-1921, "to sacrifice the Tshiokwe population in favor of the
Lunda? . . . The Tshiokwe of this territory must be brought together
under the authority of a chief of their own race." District Commissioner Labrique commented acidly: "It would have been preferable for
Mr. Coeymans to refrain from any political background work. He just
arrived at Kafakumba, has had no time to learn anything, and

already, everything is being decided—in a very bad way, unfortunately."[14] As for Tshipao, the district commissioner felt, his sous-chefferie "should disappear upon the death of the current incumbent."[15]

But while the administration had by that time made up its mind not to back Tshipao's claims to the throne, the project of organizing autonomous chefferies for the Cokwe and other non-Lunda groups had by no means been abandoned. Belgian officials had not failed to notice that all non-Lunda headmen had abstained from attending Muteba's funeral and from paying homage to his successor.[16] They were also aware, as Labrique had suggested earlier, that Kaumba's dependence on administration support would make it nearly impossible for him to block a policy of ethnic autonomy as long as official recognition had not been extended to him. A series of minor incidents between Lunda and Cokwe toward the end of 1921 provided the administration with some additional motivation and a decisive turn was reached on April 20, 1922, when Vice-Governor Rutten instructed the district commissioner to proceed immediately with the organization of "wholly independent" Cokwe units.[17] The vice-governor's choice of words may have been somewhat less than felicitous, however, and when rumors of the new policy reached the Sandoa area, there was immediate friction between the local Lunda chiefs, Mwene Matanga and Mbako, and some of the Cokwe headmen in the area, notably Samazembe, who would later be recognized as chief of the west central Cokwe. In the Kafakumba area, the leading Cokwe headman, Tshisenge Kapuma, had died in November 1921 but his nephew Sakundundu intensified his efforts to rally support from the local Cokwe population. Coincidentally, the inception of a more systematic exploitation of northeastern Angola by the Portuguese was beginning to prime a new current of Cokwe migration into the Congo. In the Dilolo region, for example, groups traveling along what had been the last active route of Cokwe trade and penetration (and would soon become the path of the railway line advancing from Lobito) asked the Belgian administrator for permission to settle in mid-1922, thus adding to the fluidity of an already complex situation and prompting several candidates to advance competing claims to leadership.[18]

In fact, what the vice-governor had in mind turned out to be something less than the complete severance of the Cokwe's subordination to the Mwaant Yaav. The Belgian administration had only just begun to discover the traditional Lunda distinction between land rights and political authority and the vice-governor's somewhat simplistic view of the matter was that "the respect of the Mwata Yamvo's land rights requires that the Tshiokwe should pay him an adequate periodical royalty—to be determined—for the transfer of the lands they occupy."[19] These views were in turn reverberated by the district commissioner to his subordinates in the field. The new policy, he explained, was based on the following considerations:[20]

a) The Tshokwe represent a distinct population which does not recognize the Mwata Yamvo for its chief;
b) the Tshokwe unanimously acknowledge that they are occupying Lunda lands;
c) both Lunda and Tshokwe keep the memory of their victories, but also of their defeats;
d) in several areas, the Tshokwe represent a majority of the population.

Accordingly, he went on, it had been determined that the Cokwe should not render to the Lunda "either the tribute or the signs of respect owed to chiefs" but that they should pay the Lunda an "annual royalty for occupying their lands." Field officials were instructed to "deal directly with the Tshokwe headmen without any Lunda intermediaries, in order to prevent the latter from giving orders or collecting tribute." The annual tithe to be paid by the Cokwe would be collected by the administrateur and turned over to the local land chiefs. At the same time, the district commissioner made it clear that there would be no question of recognizing a Cokwe paramount equivalent to the Mwaant Yaav. The Cokwe should be encouraged to regroup into homogeneous communities and the new policy "should not be forcibly introduced where no conflict is occurring."

Even when equipped with such qualifications, the new policy could hardly be expected to please the Mwaant Yaav, who threatened, somewhat gratuitously, to drive the Cokwe from his lands if they refused to give him tribute. He rejected the notion of an annual tithe, arguing (with considerable foresight, as it turned out) that "after two or three years and despite any written commitments, the Tshokwe, with their usual *mayele* (wiles) will insist that what they have paid represented a purchase price and claim to be the owners of the land."[21] Furthermore, Kaumba argued, the Cokwe headmen who were being considered for recognition by the colonial authorities were in fact impostors since all the major Cokwe chiefs, notably the Mwa Tshisenge himself, had their residences in Angola. The district commissioner retorted with a sharp rebuff:

> The Mwata Yamvo is living in a fool's paradise. The Tshiokwe are every bit as worthy of interest as the Lunda; they too are entitled to life and liberty. If only the Lunda had numerical superiority, but in many areas they are a definite minority. . . . When the Europeans arrived, the Lunda had not reached Sandoa in their reconquest of the lands they had lost. The southern portion (of the district) could thus be regarded as belonging to the Tshiokwe as a prize of war. Please tell the Mwata Yamvo that, should he persist, I shall suggest to the Vice Governor to regard as Tshiokwe lands 1) those which the Lunda had not reconquered when the Europeans arrived, 2) those who are occupied almost exclusively by the Tshiokwe, such as Mwene Ulamba and Mwene Kamwanga, in Kafakumba territory. The Mwata Yamvo is in error if he thinks that the government would assist him in driving out the Tshiokwe. This would amount to turning into a waste a land which the Lunda can neither populate nor develop.[22]

The Cokwe themselves, on the other hand, were far from unanimous in their reactions to the projected policy change. None of them was particularly enthused by the idea of making annual payments to the Lunda but the northernmost Cokwe communities insisted that any such payments should be made directly to the Mwaant Yaav, rather than to the local land chiefs. In this fashion, it was felt, any thought of their being subordinated, even indirectly, to a Lunda *cilool* would be ruled out once and for all. Among the unorganized Cokwe of the Dilolo area, however, the perspective was rather different. In that region, the local administrator explained,

> many individuals coming from Angola to settle in the Belgian Congo have been paying the Lunda headmen very small sums to occupy their lands. All these migrations have been on an individual basis, unlike in the middle portion (of Lunda) where whole clans, led by their chiefs, had conquered the land.[23]

For these southern groups, therefore, the status quo was satisfactory and, being uninterested in political centralization, they had little use for the sort of relative upgrading which a direct tributary link to the Mwaant Yaav was expected to create for the more structured Cokwe groups farther north.

By the end of 1922, the situation appeared to be almost completely deadlocked and the district commissioner reported rather somberly that "the new Lunda-Tshiokwe policy satisfies neither race. The Mwata Yamvo and the Lunda chiefs won't hear of it, while the Tshiokwe show little inclination to pay the royalty."[24] Kaumba demanded without success to be authorized to travel to Elisabethville in order to present his case personally to the vice-governor general. But the provincial authorities were now clearly intent upon settling the Lunda-Cokwe issue — once and for all, they hoped — and they pressed on with the implementation of the new policy. This took up most of the year 1923. The new district commissioner, Van den Byvang, who had succeeded Labrique at Sandoa, urged his subordinates to proceed with all determined speed, insisting that ethnic autonomy was "the only possible policy" and that "if it had not been for the distrust shown by the Tshokwe toward us, they would have been organized under their own traditional chiefs a long time ago."[25] Soon, he was able to produce results, at least on paper.

Three major issues, interwoven to some extent, were involved in the formulation of the new policy: the extent and routing of the tithes to be paid by the Cokwe; the selection of future Cokwe chiefs; and the delineation of land areas to be placed under their authority. The idea that payments to be made in lieu of tribute should be collected by Belgian officials and turned over directly to the Mwaant Yaav, rather than to local land chiefs, was accepted with relative ease by the major protagonists in the negotiation. Cokwe headmen of the Sandoa and Kafakumba regions were satisfied that this procedure would reduce to a minimum the tangible signs of Lunda paramountcy. Kaumba, despite his earlier reservations, realized that the Belgians were determined to proceed with the organization of autonomous Cokwe units and appreciated the fact that direct payments would augment his personal treasury and thus his leverage over the Lunda court circles. As for Belgian officials, they naturally found the system of a direct rebate to the Mwaant Yaav far less cumbersome to operate than the alternative procedure of turning over the Cokwe tithes to a number of land chiefs.

The second, and in many ways the central, problem was to determine which Cokwe communities would be granted autonomy and to find out which leaders, if any, they would accept. To a considerable extent, this problem was in turn linked to the question of how much land should be placed under autonomous Cokwe chiefs, an issue rendered nearly insoluble short of extensive population transfers by the almost inextricable imbrication of Lunda and non-Lunda groups. As suggested earlier by Labrique, the Belgian authorities eventually settled for the minimalist solution of recognizing as autonomous units only those areas where the population was solidly Cokwe and where explicit demands for self-government had been voiced, while encouraging those Cokwe villages that were not made part of an autonomous area to relocate in the nearest Cokwe chefferie. But since only four Cokwe headmen were eventually recognized as chiefs, those Cokwe notables who did not "make the grade" were understandably reluctant to place themselves under the jurisdiction of their more fortunate peers. As a result, even when the process of ethnic restructuring had been completed (or, more accurately, after the administration had reached the limits of its attention span on this issue), there remained large numbers of Cokwe under the authority of Lunda chiefs — sometimes amounting to a majority of the population, as in the case of Mwene Ulamba, Kayembe Mukulu, Tshisangama, Muteba, Mbako, and others.[26]

For their part, would-be Cokwe chiefs naturally tried to round up potential subjects, not only by seeking recognition of their authority from local groups but also by inviting remote sections of their clans to come and settle under their protection. In the east central portion of Lunda, the future Chief Sakundundu attempted to recruit immigrants from the right bank of the Kasai River, over one hundred miles away, and from the southeastern corner of the Mwaant Yaav's chefferie.[27] Over the years, the Cokwe had settled heavily along the trade routes to Mutombo Mukulu and Kasongo Nyembo, creating a Cokwe corridor between the lands of the Mwaant Yaav and those of the Kayembe Mukulu. Tension between these two chiefs, which had been manifested in 1920 by Muteba's failure to impose his candidate to the succession of Kayembe Mukulu, had also taken the form of a conflict of influence in the buffer zone of Mwene Ulamba. During the Lunda-Cokwe wars, a Cokwe headman had been granted land in the north of this area by Mwaant Yaav Mushidi as a reward for saving Mushidi's life.[28] Later, his successor, Mbangu, had infiltrated the lands of Kayembe Mukulu, and the result had been a protracted tension in which each side produced a rival *mwaantaangaand* claiming the title of "Mwene Ulamba."[29]

Since the area was predominantly Cokwe in terms of population, Sakundundu could hope to benefit from the animosity between the two contestants in order to extend his authority over all Cokwe groups. But Mbangu II, the successor of the Cokwe headman who had saved Mushidi's life, was hoping to be recognized as an autonomous chief and showed no inclination to throw in his lot with Sakundundu. As a result, the east central Cokwe chefferie that was eventually organized under Sakundundu did not extend northward to the limits of the district, and a Lunda was finally recognized as subchief over an area where the population was 70 percent Cokwe. As for Mbangu II, he and his successor continued to agitate in vain until the early 1930s to have their own chefferie and even managed to enlist the Mwaant Yaav's support for their claims, but the Belgian administration refused to re-open the issue of Cokwe claims, which it considered "closed once and for all."[30]

In the solidly Cokwe region extending from Sandoa westward to the Kasai River, the local administrateur de territoire bolstered the position of the senior Cokwe headman (Samazembe, of the Tshisenge clan) by instructing Lunda chiefs, such as Muteba, to abstain from any further dealings with the Cokwe residents of their chefferie and by resorting to the old practice (once used on behalf of the Mwaant Yaav) of touring the villages in the company of the prospective chief. On one such tour, sixty-six village chiefs came forth more or less spontaneously to offer tribute to Samazembe.[31] Farther to the southeast, however — and particularly between the rivers Lulua, Kasileshi, and Lueo — a strong concentration of Cokwe villages of the Kandala clan remained unorganized. A senior member of that clan, who had moved into the Congo from Angola in 1922 in the hope of being recognized as chief, had settled to the east of Dilolo but showed as yet no promise of being accepted by any local groups. Not only did a substantial portion of the Cokwe in the Dilolo region belong to the Tshisenge rather than to the Kandala clan, but also patterns of residence were so fragmented that the local Belgian official, while paying lip service to the new policy, estimated that, if any attempt were made to organize ethnically homogeneous chefferies, substantial numbers of Cokwe and Lunda would be forced out of the lands where they had been living for years.[32] The district commissioner's sweeping reply was indicative of the administration's desire to settle the Lunda issue as expeditiously as possible, even if this required a certain amount of arm-twisting:

> The Tshiokwe should gather along family and clan lines; clan and family heads are naturally indicated to convince their own people; as for us, we cannot proceed with these population transfers on our own, i.e., without the consent of the interested parties. On the other hand, however, if the Tshiokwe will not submit to the customs of the Lunda whose land they might have occupied only recently, we have every right to expel them from those lands upon request from the Lunda. These remarks must be brought to the attention of the Tshiokwe population for it is quite certain that, once the delineation of the lands belonging to Lunda chiefs is completed everywhere, those Tshiokwe who insist upon staying will have to bear all the drudgery on the part of the Lunda; their refusal to comply will force them to leave.[33]

The delineation of the lands to be surrendered to the Cokwe was by far the most delicate problem, and, no matter how anxious the colonial authorities were to stabilize the different ethnic groups, it would take several years before a final demarcation could be achieved. As mentioned earlier, many Cokwe villages had reached an acceptable *modus vivendi* with the Lunda land chiefs and were not particularly attracted by the prospect of living under Cokwe chiefs, especially if it meant relocating their village in an unfamiliar area. A few, such as Mbangu II of Mwene Ulamba or Mwa Koji of Tshipao (later Tshisangama), had even been recognized as land chiefs by the Mwaant Yaav and paid tribute directly to the Musuumb. Many of these headmen were in fact hostile to more recent Cokwe immigrants. On the other hand, the prospect of supervising massive transfers of population was highly unattractive to the administration. Furthermore, most local Lunda chiefs vigorously opposed the idea of transferring land rights to the Cokwe. "If we were to believe the Lunda notables," wrote the administrateur for Sandoa, "they would occupy the land in its entirety and recognize no rights whatsoever to people who have been settled for years, sometimes for generations, and who are far more numerous than they are."[34] These factors, combined with the administration's desire to reconcile minimum disruption with maximum rapidity in settling the Lunda-Cokwe problem, led to the signing of the so-called Kapanga Convention on September 29, 1923.[35]

The Kapanga Convention was ratified, on the Lunda side, by Mwaant Yaav Kaumba surrounded by his entire council and, on the side of the colonial administration, by District Commissioner Van den Byvang and by the administrateur de territoire stationed at Kapanga, under whose jurisdiction the Mwaant Yaav was technically placed. It outlined those portions of the Lunda heartland that would be turned over to the Cokwe; no further such transfers ever took place after that date. The actual demarcation was left to be settled by future protocols, but the overall pattern of the Cokwe domain emerged clearly from the agreement: one area situated between Sandoa and the Kasai River (corresponding to the modern Samutoma chefferie); an east central zone, initially scheduled to include the lands of Sakundundu as well as those of Mbangu (the northern half of the modern chefferie of Mwene Ulamba) but subsequently restricted to the first of these two units because of Mbangu's insistence on a separate chefferie; and finally a much smaller area in the south extending eastward from Dilolo and corresponding to the modern chefferies of Tshisenge and Kandala.

The agreement further provided that no person would be forced to leave his current residence against his will, "under the absolute condition of submitting to the orders of the chief of his residence." The Mwaant Yaav agreed to cooperate to ensure a swift relocation of those populations that wanted to move their residence, with the provision that they would be given time to plant new crops and would retain the right to reap all standing crops at the site of their former settlement.

But the most important part of the convention in the eyes of the Lunda court — and, as it turned out, the most fragile — was the clause which specified that an annual royalty of no less than twenty centimes per taxpayer would be collected by the Cokwe chiefs and turned over to the Mwaant Yaav through the intermediary of the Belgian administration. From the outset, the nature and significance of this annual tithe were the objects of various misconceptions, some of them deliberate. According to the agreement, the rate of payment had been determined "by taking into account the tax rebate previously paid to the Lunda chief and the occasional tribute he received from Tshiokwe notables." Thus, in the eyes of the Belgian administration, only an unspecified portion of the proposed annuity (in fact a relatively small one) had any kind of traditional basis — and that was referred to as an "occasional tribute" rather than a well established custom. Kaumba and his entourage, however, chose to interpret the annuity purely as tribute, and thus as a sign of the Mwaant Yaav's continued suzerainty.

Despite the fact that the money collected by Cokwe chiefs would not be carried directly to the Musuumb but turned over to the administrateur de territoire, then centralized by the district commissioner and made payable to the Mwaant Yaav in the same way as his regular salary, the Cokwe by and large viewed the whole procedure in the same light as the Lunda, and reacted accordingly. Mwa Tshisenge (Samazembe) of Sandoa accepted the Kapanga Convention when it was presented to him three weeks later, but, the district commissioner admitted, "he did not appear to appreciate very much the royalty which his people will have to continue paying to the Lunda paramount."[36] From the very beginning, the district commissioner recognized that "a point which, in my opinion, is bound to raise some difficulties is the payment of a fixed royalty by the Tshiokwe." He went on to say:

> Although it is equitable that this royalty should be paid, and although the amount is minimal, the natives cannot understand why, now that they are free, they should still give a royalty to the Mwata Yamvo for any reason whatsoever, and they persist in interpreting this payment as an act of submission, since it has the characteristics of an actual tribute in their eyes.[37]

The solution to this delicate problem, he suggested, would be for the colonial administration "to assume the payment of this royalty by withholding the amount from the returns of the native head tax collected among the now autonomous Tshiokwe populations."[38] The provincial authorities went one step further and invited the district commissioner to study the means of consolidating the annuities into a lump sum to be ammortized over one or several years, so as to eliminate once and for all the possibility of future Lunda claims on the lands granted to the Cokwe by the 1923 agreement.[39] These instructions were reiterated in 1926, but as will be shown below, District Commissioner Van den Byvang had in the meantime developed serious misgivings regarding the wisdom of the policy of ethnic autonomy. These doubts were shared more or less explicitly by his two successors, and thus it was not until 1932 that the annuity was finally discontinued.[40]

Other aspects of the new policy were equally difficult to implement. Apart from some of the conflicts mentioned earlier, the problem of demarcating the limits of the Cokwe zones was a relatively simple one, except in the southwest where, in addition to the interpenetration of several ethnic groups—Lunda, Cokwe, Ndembu, Alwena, etc.—the generalized use of the Cokwe language confused and baffled officials. We can appreciate their perplexity when faced with the Lunda dignitaries' claim that the lands situated between the Cokwe nucleus in the Dilolo area and the territory of the Alwena (the future Dumba chefferie) were populated by "Cokwe-speaking Lunda."[41] A comparable situation prevailed at Saluseke, between Dilolo and the Kasai River. To make things worse, most of the inhabitants of the Dilolo area were hostile to any form of organization. "The term of anarchy," wrote the exasperated district commissioner, "is not too strong to describe the political life of this area . . . . It would be tempting to try direct administration, but the government is against it."[42]

A more serious problem was the lack of authority and generally low caliber (in Belgian eyes) of the Cokwe candidates to a chiefly position. Samazembe of Sandoa, whose ambition was to be recognized as the senior Cokwe chief in the Congo (an aspiration probably inspired by the example of the Mwaant Yaav) and who, for that purpose, claimed, without any real foundation, the prestigious title of "Mwa Tshisenge," feigned a trip to Angola, purportedly to receive from the real Mwa Tshisenge the insignia of office, and reappeared with an assortment of spurious insignia. Questioned by a skeptical district commissioner, he confessed his fraud but, for lack of any other suitable candidate, was nevertheless invested and permitted to make de facto use of the title of Mwa Tshisenge, much to the annoyance of local Lunda chiefs.[43] Thereafter, the administration's opinion of Samazembe declined steadily, and within three years the district commissioner frankly admitted that Samazembe was hopeless (from a European viewpoint) and that only his eventual replacement by his more dynamic brother, Samutoma, could set the chefferie on the right track.[44]

Samazembe never desisted from his early claims to be recognized as the Cokwe counterpart to the Mwaant Yaav, but without any real success. Apart from the questionable basis of his traditional authority, the self-proclaimed "Mwa Tshisenge" simply could not match the flair for decorum cultivated by Lunda court circles. Thus, when the Mwaant Yaav and his retainers traveled to Sandoa during the latter part of 1926, the mortified Samazembe could not prevent some of the Cokwe headmen under his jurisdiction from paying their respects to the Lunda paramount. Two years later he refused to attend a conference of chiefs convened by the district commissioner as a prelude to the attempted creation of a pluri-ethnic native administration structure, because the conference was to be presided over by the Mwaant Yaav. During the centennial celebrations of Belgian independence in 1930, in the apparent desire to upstage the Lunda paramount, Samazembe deliberately made his entry at Sandoa after the Mwaant Yaav's arrival, to the accompaniment of an impressive array of drums and gongs—for which attitude he was fined the rather stiff sum of two hundred francs.[45]

The administration's opinion of Sakundundu, of the east central Cokwe group, was hardly more favorable, at least initially. Less than one year after the signing of the Kapanga Convention, he was being described as "not very serious for the job"; and a few months later as "incapable of listening to reason," and as "thinking only of the personal benefits to be derived from his position." Still another report dismissed him as "an erotic monomaniac."[46] After a while, however, the colonial administration came to regard Sakundundu as reasonably satisfactory, but his customary status remained somewhat ambiguous. A member of the Tshitanga clan, he was the nephew of "Tshisenge" Kapuma, who had attempted to establish his authority over the east central group but had been driven out as a result of repeated clashes with the Mwaant Yaav. The real head of the Tshitanga clan, however, one Samuhungu, migrated from Angola in 1926 and settled near Kafakumba, but without attempting to supplant Sakundundu in his role of go-between with the European authorities.[47]

In the Dilolo area the situation was, if possible, more tangled still. The local Cokwe population belonged predominantly to the Tshisenge and Kandala clans, but neither group had a recognized head, and the several aspiring chiefs (two of whom at least were "carpetbaggers" who had moved into the area in the hope of securing Belgian recognition) vied against one another with such gusto that Belgian officials, in desperation, briefly lumped all Cokwe into a secteur. This decision was soon reversed, however, and two chiefs—"Mwa Tshisenge" Saienge and "Mwa Kandala" Mashata—were officially recognized between 1925 and 1926 despite the misgivings felt by some European administrators. Chief Saienge, an old man who was almost totally ignored by the local population, died in January 1926, however, and the Mwa Kandala had to be arrested in 1927 for inciting the murder of a Lunda headman who had crossed his lands in a *tipoy,* the litter usually reserved for chiefs. Once more, the Belgian administration reverted to the practice of using a single council of notables to govern the two chefferies. "To rule this tribe is quite a task!" commented the administrateur from Dilolo, to which the district commissioner added:

> We will finally have to admit that the Tshiokwe have never had any real political cohesiveness and that whenever they have been forced to obey, they only chafed under restraint. Under the circumstances . . . we must again try to give them a customary chief—an able and competent one, of course—and if we fail again this time, we should not hesitate to force upon them a chief of our own choice.[48]

Two new chiefs were eventually recognized, but problems continued. Saienge's successor had to be deposed in October 1934 while the original Mwa Kandala, after serving his prison term, recaptured much of his former authority and became a sort of occult chief.

The local administration's gradual disenchantment with the Cokwe was matched after 1925 by an increasing admiration for Lunda institutions and by a growing skepticism toward the policy of ethnic autonomy initiated only a few years earlier. In the opening portion of a study written in 1926, shortly before his transfer to the Kwango, District Commissioner Van den Byvang stated flatly: "Sincerity compels me to declare that it was a major error on our part to have completely detached from the authority of the Mwata-Yamvo the Tutshiokwe who lived on Uluunda lands and recognized its chiefs."[49] And in his closing report as district commissioner for Lulua, Van den Byvang invoked the "racial affinities" between the Lunda, Cokwe, and Alwena and argued strongly in favor of a form of native administration that would institutionalize the subordination of the latter two groups to the Mwaant Yaav.[50] The vice-governor declared himself unimpressed by Van den Byvang's evidence, however, concluding that the decision in favor of ethnic autonomy for the Cokwe and Alwena was "immutable" and that the issue was closed "without any possible reopening."[51]

Van den Byvang's successor, Caroli, avoided clashing with his superiors and proceeded with the long-delayed organization of autonomous Lwena chefferies, but he displayed no particular zeal to satisfy the provincial authorities' repeated request to have the Cokwe annuity to the Mwaant Yaav converted into a lump payment. His appraisal of the native administration program for the district in early 1927 was certainly less than enthusiastic:

If I may use the expression, I would compare the work we did to the launching of a ship. I would question neither the competence nor the efforts that went into the construction itself but, in my opinion, the final operation, the test of its "seaworthiness," was a failure. Had we checked, we would have discovered that some elements of the machinery lacked the proper finish and that others perhaps were imperfectly adapted to the whole set of gears.[52]

Derogatory judgments regarding the performance of Cokwe chiefs were frequently contrasted, implicitly or explicitly, with the supposedly innate qualities of leadership to be found among the Lunda. "They have a fine bearing and a lot of ability," wrote a district commissioner in 1931, "while the Tutshokwe lack style and look like the descendants of mere gang leaders."[53] To this stereotype, which flourished at the district level and below, corresponded a counter-stereotype found mainly among provincial officials which saw the Lunda as "a degenerate race without much spunk,"[54] and the Mwaant Yaav as "a chief whose political power was falling apart when we arrived in the land and who derives the semblance of authority he wields from European officials."[55] The same circles also tended to be more tolerant of the "faults" of the Cokwe chiefs, which they felt were common to many of their peers.[56]

The obstacles encountered in implementing the policy of Cokwe autonomy during the early nineteen-twenties had their repercussions on the status of the other major non-Lunda group. The Alwena of the southwest fringe of Katanga had escaped previous attempts by the administration to organize them. Indeed, as will be remembered, SaTshilembe, the head of the Katende clan in the Congo, had elected to move to Angola in 1918 rather than accept subordination to the Mwaant Yaav. Since that time the Belgians had repeatedly tried to entice him back to the Congo with the tacit approval of the vice-governor, who tolerated this one exception to the implicit agreement among colonial powers not to "steal" one another's subjects.[57] Meanwhile, however, another Lwena group, that of Tshilemo, had agreed to recognize the Mwaant Yaav's suzerainty. In the latter part of 1925, old SaTshilembe, who had apparently had difficulties with the Portuguese authorities, returned to the Congo, and District Commissioner Van den Byvang at once announced his intention of placing the entire Lwena population under the Mwaant Yaav's authority.[58]

The decision probably reflected to no small extent Van den

Byvang's growing doubts about the wisdom of organizing autonomous Cokwe communities, but, coming as it did so soon after the Kapanga Convention of 1923, it appeared to hold the threat of a reversal of the policy of ethnic autonomy inaugurated after World War I. The reaction from Elisabethville was that the directives concerning Lunda-Cokwe relations should equally be applied to the Alwena. "We know from a study prepared in 1913," wrote Vice-Governor Bureau, "that already at that time [the Alwena] were independent. Thus, it seems to me that attempting to subordinate the Alwena who have been independent for at least thirteen years to a chief whose authority applies only to the extent that we support him is an operation involving more risks than advantages."[59]

Van den Byvang's successor to the post of district commissioner proceeded to comply with the vice-governor's instructions, only to discover that another branch of the Katende clan farther to the east was also demanding its autonomy. To complicate matters further, the head of the Tshilemo group, who had accepted the Mwaant Yaav's authority all along, refused to have anything to do with the Katende and let it be known that he would not settle for anything less than what his neighbors were being granted.[60] Once more, the instructions from Elisabethville were that the Alwena should be made independent, even if it meant recognizing two chiefs.[61] In fact, the final paragraph of this episode was not written until some three years later. The transfer to Angola in 1928 of the area known as the "botte de Dilolo" entailed the loss of approximately one-half of the proposed Katende chefferie. The death of SaTshilembe in early 1929 further contributed to the fluidity of the situation among the Alwena, and thus it was not until 1930 that a new Chief Katende was finally recognized officially. In time, Tshilemo and the eastern branch of the Katende clan were also organized as separate chefferies.

The position of the Lunda aristocracy against this background of contradictory attitudes was understandably ambiguous, and their reactions were based on the cautious tactical exploitation of favorable circumstances rather than upon any discernible long-term strategy, excepting perhaps that of survival. Kaumba, who had just been ritually invested at the time of the Kapanga Convention, immediately after signing that document was faced with the stunned resentment of many Lunda tributary chiefs who, unlike the court dignitaries, were to suffer a loss of income through the removal of Cokwe communities from their jurisdiction and who denounced him bitterly as the first Mwaant Yaav ever to surrender portions of his domains without a fight.[62]

Against such recriminations the Mwaant Yaav's main trump card, if not his only one, was his traditional prestige, both in the eyes of the Africans and in those of the colonial administration—or, more accurately, his ability to transmit legitimacy to the authority of Lunda and even non-Lunda headmen. The Kapanga Convention inaugurated a pattern according to which a community's autonomy was paradoxically signaled by the establishment of direct tributary links to the Mwaant Yaav. Under the circumstances, and knowing full well that the Lunda paramount was in no position to intervene in the day-to-day administration of local communities, many headmen throughout Lunda sought to achieve for themselves the position of direct tributaries. Not only did this factor enable the Lunda paramount to play off one notable against another; it also contributed to reinforce in Belgian eyes the belief that the Mwaant Yaav's influence reached into the remotest corner of Lunda, thus causing him to be called upon to arbitrate succession problems or land disputes—which in turn enhanced the value of his patronage and increased his potential leverage.

Such manipulations, which Kaumba had originally employed against Tshipao's supporters following his election, had their obvious limits, namely the Belgian administration's reluctance to disrupt existing situations when these worked to their satisfaction. "The Mwata Yamvo's tendency to support and even to ratify the pretensions of some dissidents vis-à-vis a recognized chief or sub-chief," wrote the district commissioner in 1927, " . . . . nearly always turns against his prestige and authority in the end, as we can never reopen issues that have already been settled."[63] Even though local situations may have been maneuvered by the Mwaant Yaav more often than Belgian officials realized or were willing to admit, it is true that a chief who performed what the colonial authorities expected of him—in terms of road-building, manpower recruitment, tax collection, cooperation with the missions, etc.—could count on their support against local dissidents, whose claims merely served the administration to remind the chief that he was not indispensible. Thus, in a sense, the Mwaant Yaav's attempts to manipulate the Lunda aristocracy were bound to increase that chief's subservience to the colonial system, at least indirectly. Indeed, Kaumba's own leverage on the notables was, in the final analysis, predicated upon his ability to maintain good relations with the colonial administration.

During an initial period of a year or so following the signing of the Kapanga Convention, the Mwaant Yaav, whether out of personal resentment for having been browbeaten into signing the agreement, or because he had been criticized by his entourage for being too docile, was distinctly noncooperative, and the resident Belgian administrator at Kapanga even suggested that Kaumba's replacement might be in order.[64] The possibility was not seriously considered, but District Commissioner Van den Byvang nevertheless felt that it was necessary to "demand of the Kaluunda chief tokens of respect and obedience toward any European official" and that the Mwaant Yaav could be highly useful if "trained to our methods, well advised, and controlled unobtrusively."[65] At the same time, Kaumba's claims on the Kete borderland area were being denied and provincial authorities were urging the discontinuance of the Cokwe annuity. Under the circumstances, one could hardly expect the Mwaant Yaav to show a burning zeal to help the colonial administration, especially regarding matters outside the traditional process. Yet, as illustrated by the issue of African manpower recruitment, this is precisely what Belgian officials expected of him as proof that he had embraced the ideal of "modernity."

Over the previous decade, chiefs throughout Katanga had been consistently urged to encourage their subjects to sign up as laborers in the fast-developing mines of the southeast. The major firms of the industrial region of Upper Katanga had joined forces to organize a recruiting agency, the Bourse du Travail du Katanga BTK; later renamed Office Central du Travail au Katanga: OCTK), to which provincial authorities had promised the full cooperation of all field officers. The result was a brutal process which soon generated serious concern among some missionary and administration circles but was not effectively altered until the depression, when massive short-term recruitment was gradually superseded by a manpower stabilization program.

In Lunda, where all-weather roads had been unknown before 1920 and where the first "tin Lizzies" appeared only in 1922, the first significant contingent of African laborers had not been recruited until 1920, and then mostly from the south,[66] but as the demand for manpower escalated, Belgian officials were under increasing pressure to tap local resources. Much to their annoyance, however, Kaumba was distinctly uncooperative. During the first half of 1925, for example, when total recruitment from the district by the BTK exceeded one thousand, no more than eighty men were recruited from Kapanga territory, and District Commissioner Van den Byvang noted irritably:

> The Mwata-Yamvo does not understand that his administrative pre-occupations should become modern. What, for instance, has he done about the alarming problem of manpower, where everything has been attempted to make him understand our difficulties? He has taken no interest in the problem and, while promising me his un-stinting cooperation during his visit to Sandoa, he told other persons, as soon as he had left, that he would pay no attention to my entreaties! Yet, it would not have been much of an imposition, considering that there are — as I have since learned — over one thousand laborers from Kapanga territory in the industrial areas of Upper Luapula, almost all of them voluntary recruits, while I was asking the Mwata-Yamvo for only six hundred! . . . The Mwata Yamvo should help us here also: (1) by methodically channeling this man-power instead of letting it go to waste on short-term contracts; (2) by sending this manpower to the Upper Luapula area where the short-age is much more intense.[67]

Meanwhile, however, the Mwaant Yaav had found a champion in the person of a young Belgian official named Léon Duysters. Duysters, who took up his post as administrateur de territoire for Kapanga on September 9, 1925, had acquired from the lectures of Georges Van der Kerken at the Antwerp Colonial Institute an admiration for men like Lugard and Lyautey and a belief in the principles of indirect rule. He soon established excellent relations with Kaumba, and his first reports reflected his appreciation of what he called the Mwaant Yaav's "tre-mendous prestige."

> His every word, his every gesture, is obeyed. Every native feels to-ward the Mwata Yamvo a respect which I would dare to call divine; his person is sacred. The same goes for his ministers and members of the Lunda grand council. The term of chief is not adequate for the Mwata-Yamvo: he is a veritable king surrounded by his ministers and by the great dignitaries of his court . . .
>
> It seems that the Mwata-Yamvo fears that the government — which, in his eyes, pursues a capricious policy — will detach the out-lying areas of his kingdom piece by piece . . . . Under the circum-stances, it is not surprising that the Mwata-Yamvo's assistance to the

administration is reduced to a bare minimum when he is residing at
Musumba.[68]

Yet, Duysters added in a subsequent report:

> The Mwata-Yamvo knows full well that his fate and his future lie in
> our hands, but his position is often difficult: he is an elected Chief
> and must retain his popularity. But our interest is sometimes con-
> trary to the interest of the natives, and the most striking example is
> precisely that of labor recruitment for which he showed no interest.
> The Mwata-Yamvo did not supply the 600 laborers requested of him
> by the B.T.K. because that agency is very unpopular with the
> masses and because, had he used his authority, a part of this un-
> popularity would have reverberated against him.
>   In my humble opinion, it is a serious mistake, anyway, to have the
> chiefs intervene in the question of labor recruitment. The real in-
> terest of the natives is not for them to be turned into industrial wage
> earners, but into good agricultural producers who will stay in their
> villages amidst their own people. The important problem of labor
> recruitment can be solved without sacrificing the vitality of the
> already sparse population of the colony.[69]

By far the most unorthodox of Duysters' ideas, however, was his
suggestion that Lunda should be disjoined from the native administration system and organized as a protectorate.

> The Mwata-Yamvo is the head of a native State which extends to the
> sources of the Lulaba and of the Zambezi, from the Kasai to the
> Lubilash; that state has maintained itself and exists. It would be a
> regrettable political error not to adopt the one effective measure
> *that would guarantee the future of this state and the stability of its
> institutions,* namely, the official recognition of the Uluunda King-
> dom and its organization in the form of a Protectorate, including
> the Tutshiokwe and Aluena components, blood brothers of the
> Aluunda conquerors who have infiltrated among them, are still im-
> migrating at the present time, who all claim to be children of Lueji,
> their common ancestor, and who, save for an occasional exception,
> recognize the Mwata-Yamvo as their Paramount Chief.[70]

Duysters' views were met by a barrage of objections at the district
and province levels. Despite his own growing conviction that Cokwe
and Lwena chiefs should not have been made autonomous, District
Commissioner Van den Byvang commented, "our youthful official

seems to be subjugated by the coterie of . . . the Uluunda Council and their spokesman the Mwata-Yamvo."[71] "I agree with you," concurred the vice-governor, "that the administrator for Kapanga, carried away by his optimism and his enthusiasm, deludes himself regarding the Mwata-Yamvo's actual power . . . . He obviously lacks experience and level-headedness . . .[If he was] better informed, he would know that for many years we have tried in vain to persuade Mwata-Yamvo to lend us effective assistance."[72]

More concretely, these same officials objected, Lunda was too underdeveloped to be anything but a "mendicant protectorate," and the Mwaant Yaav was incapable of "exercising the prerogatives of sovereignty, of ensuring the respect of native customary law and of giving the European administration a genuinely active assistance."[73]

In fact, the real dispute was one between conflicting theories of native administration. Louis Franck's concepts of indirect rule, however ambiguous they may have been, had never met with more than lukewarm acceptance in Belgian colonial circles and had engendered some highly vocal opposition in various quarters, notably in Katanga. The reasons for such opposition are reasonably well established and are by no means unique to Belgian Africa: Kenya and the Rhodesias were at that time offering similar resistance to Lugard's ideas, and indirect rule never took root in those sections of British Africa.[74] The more specifically Latin factors of cultural imperialism and institutional rationalism were also found in the ideological arsenal of the Belgian opponents of indirect rule. In a sense, the arguments used against indirect rule, whether in the Congo or elsewhere, were often the logical extrapolation of the very same reasons that were adduced in defense of Lugard's views.

Indirect rule in Africa had been developed and had met its greatest success in areas where, for military or economic reasons, it had been found imperative to rule vast, densely populated areas with a minimal number of European officials and where relatively cohesive and centralized forms of traditional government had been encountered. Wherever rapid social and economic change was generated by European agencies, on the other hand, indirect rule usually proved unworkable or, to be more accurate, undesirable in the eyes of the various colonial interest groups, whether settlers, missions, or business firms. For each one of these groups, the existence of large, dynamic native states involved potential problems. To settlers and businessmen who sought cheap manpower and political quiescence, the problem was largely circumstantial; a chief could be "good" or "bad," depending on his willingness to cooperate, and traditional values were alternatively regarded as inferior — when compared with their Western counterparts — or commendable — when contrasted with those of the "uppity" city dwellers. Ideological commitment to direct rule was infrequent among such circles, but the very nature of their preoccupations made them fundamentally intolerant of any native administration system that would permit African chiefs to stand in the way of their interests.

The attitudes of the missions (specifically, in the case of the Congo, the Catholic missions) were at the same time more complex and more intrinsically opposed to indirect rule. "It is undeniable," concedes M. Crawford Young, "that many missionaries viewed most traditional structures as being so linked by their ritual sanctions with practices incompatible with Christianity that their disappearance was a necessary prerequisite for fulfillment of the evangelical task."[75] Indeed, the most articulate critics of Louis Franck's ideas had been missionaries. A meeting of the heads of all Catholic missionary communities held in October 1923 at Stanleyville resulted in preparation of a document which denounced indirect rule in no uncertain terms:

> The theories are brilliant, but frequently they are nothing but a lure, an infatuation with words, a facile propaganda. A colonizer who is consistent to the end in defending respect for tradition is condemned to cease colonizing . . . . In theory of course, indirect rule which, by definition, does not destroy what is deserving of respect, should be preferred . . . . But our brilliant theoreticians do not teach us how to achieve this beautiful government and cannot raise from the ground the corps of elite officials who will have the charge of guiding and controlling the harmonious evolution of native societies . . . .
>
> The Belgian Congo's official native policy [inculcates] respect for customs and for the traditional authority of chiefs. But in order to proclaim this principle, certain things are being forgotten and others are being ignored.
>
> Forgotten is the Colonial Charter, which saps the base of the chief's traditional authority; for that authority is based, on the one hand, upon domestic slavery and polygamy and, on the other, upon the occult and mysterious power that he holds from his ancestors and which he uses, either directly or through diviners or witch-

doctors, to maintain all his subjects within the narrow framework of custom; yet for these subjects the Colonial Charter proclaims freedom of conscience, of worship, and other precious liberties. Forgotten is the bulk of current legislation which limits, curtails, or destroys [traditional] authority and custom on a thousand points. Ignored are the thousands of factors which for the past forty years have been sapping and destroying authority and custom . . . . Ignored and forgotten is the fact that chiefs are fundamentally selfish . . . . Ignored again is the fact that chiefs are necessarily, as a result of custom, enemies of the State.[76]

In Katanga itself indirect rule had a formidable opponent in the person of Monsignor Jean Félix de Hemptinne, apostolic vicar of Elisabethville, and a man who never minced words. His influence was particularly strong in the Katanga section of the Commission for the Protection of Natives, an agency originally devised by Leopold II to appease his critics and which, after World War I, acted as an intermittent conscience for the colonial authorities. Reflecting the predominance of missionary elements amidst its ranks, the commission expressed serious reservations about indirect rule: "Native institutions which are merely excusable or tolerable cannot claim a protection which would assure their perpetuation," the commission flatly stated in 1923. "They are destined to disappear."[77] On the subject of native administration policy, the Katanga subcommission was even more explicit. It declared in 1925:

One wonders if the efforts are demanded of the field administration to maintain the domination of a chief whose power is crumbling do not at times exceed those that would be required for the direct administration of small clans, a form of organization commensurate with native mentality and through which European influence can be exercised successfully.[78]

In Katanga, where business and missionary circles were perhaps more influential than anywhere else in the Congo, such views found a sympathetic audience at the provincial headquarters. Vice-Governor Bureau's paradoxical view of indirect rule in the case of Lunda was that it was only conceivable after the Africans had been "educated" through a long period of direct administration:

A task of prior reorganization must be undertaken: we must educate the Mwata-Yamvo, give him a clear idea of his mission and of the responsibilities he assumes in the administration of his people, teach him to place the general good above his own personal interest . . . . The natives must in addition be able to understand the regime that we want to introduce and capable of accepting it. Since these conditions are still far from being met, it is not yet appropriate to move in the direction of indirect rule.[79]

Even though Van den Byvang, who had come to hold views somewhat similar to his own on the need to maintain the Mwaant Yaav's authority, had now left the district, Duysters still held his ground. When he turned over the administration of Kapanga territory to his successor before going on leave in July 1927, he must have known that he would not be reassigned to the area. At any rate, his final report sounds like a parting salvo:

> They are a race of chiefs, these remarkable Aluunda chiefs, who stand out by their demeanor, their dignity, their prestige as much as by their spirit of justice, their jealous preservation of prerogative, their attachment to their community and customs . . . . And the qualities of these chiefs, their admirable political organization, are serious guarantees for the future, if the government is willing to give up all notions of direct rule and to recognize this native state . . . . The Decree on Chefferies was not made to apply to such a native kingdom with its strongly hierarchized and centralized power: a Protectorate covering the entire Uluunda kingdom is the only form of government to introduce in this area. I regard the partition of the Uluunda State as an error — partition into Uluunda and Utshiokwe areas, partition between territoires and districts, partition into chefferies and sous-chefferies . . . . This is a policy of "divide and rule"; this is an encouragement to dismember the largest Native State in the Belgian Congo . . . . The Decree [of 1910] should be repealed and replaced with new provisions establishing a Protectorate in that part of the colony . . . . That Decree was inspired in part by old theories now fallen into disrepute, notably the theory of assimilation and direct rule, whose implementation has been disastrous for native policy. At the present time, these theories hardly find any defenders among colonial nations.[80]

But the last word remained of course with Duysters' superiors. Commenting ironically on the final report by the man "who probably dreamed more than once of some day becoming the Resident of an Uluunda Protectorate," District Commissioner Caroli noted that "his

'modernistic' ideas concerning our policy might cause us some difficulties and even become dangerous if they were to be shared by other
young officials."[81] From Elisabethville, the acting commissioner
general wrote to the governor general at Boma that Duysters' return to
Kapanga "was wholly undesirable" because of "his Utopian tendencies
and his lack of discipline."[82]

As Duysters left Kapanga, however, a new assistant district commissioner was appointed to serve under Caroli. Victor Vermeulen had
served from 1923 to 1926 at Wamba in the Eastern Province, an area
where ethnic fragmentation was so widespread that secteurs had had
to be resorted to systematically.[83] Upon his first inspection trip to
Kapanga during the latter part of 1927, he found the Mwaant Yaav
concerned about the gradual erosion of his authority, as had Duysters,
and offered to convene a council of Lunda chiefs. Kaumba was understandably quite favorable to this idea.[84] Vermeulen's ideas about his
new district took shape over the next few months. In early 1928 the
governor general sought the opinion of all district commissioners
throughout the Congo regarding the desirability of legalizing secteurs
(which had been resorted to fairly extensively since Louis Franck's
original circular of 1920 but had never had a statutory existence) by
amending the Decree of 1910. Vermeulen took over from Caroli as
district commissioner in March 1928, just in time to reply to the consultation with the unexpected suggestion to organize a vast "Lunda-
Tshiokwe Sector," covering virtually the entire district and divided
into six "subsectors" corresponding to the six territoires.[85] He developed the idea in subsequent reports: the organization would involve, he
submitted, a three-tiered council system, with a council in each chefferie or sous-chefferie, one council for each "subsector" (i.e., territoire) which "could be presided over by a dignitary from the Musumba
or by a chief deputized by the Mwata-Yamvo," and, crowning the
whole structure, a council for the entire sector. Such an organization,
Vermeulen argued, would reinforce the Mwaant Yaav's authority over
the chiefs who formally recognized his paramountcy but frequently
tended to act independently and to deal directly with Belgian officials.
Thus the tendency of the Lunda court to support local headmen
against the chiefs recognized by the colonial administration would in
turn lose its raison d'être.[86]

The reaction of provincial authorities was one of circumspect, con-
ditional approval. This was more than the new district commissioner

needed to proceed with his plans. A first council presided over by Kaumba was held at Sandoa in December 1928 and was attended by a number of chiefs from various parts of the district, including Sakundundu, Kazembe, and the successor of the late Tshipao. Subsequently, similar councils were convened more or less regularly in each territoire.

In fact, the organization of a Lunda-Cokwe sector (which, incidentally, was meant also to include the Alwena) was only one facet of Vermeulen's ambitious plans for the district. The reform of native administration was to be completed by the parallel development of native tribunals and native treasuries. Here, the district commissioner stood on firmer ground; the creation of these agencies was based on explicit government instructions—a possible reason why provincial authorities reserved their judgment on Vermeulen's other initiatives. A decree of April 15, 1926, had established a legal framework within which native jurisdictions could be organized, and a provincial ordinance of May 30, 1927, had proceeded to recognize a number of native tribunals in Katanga. In Lunda itself Duysters had even attempted to organize a native tribunal under the Mwaant Yaav as early as November 1925. Now, despite the misgivings felt by many European magistrates, the development of these jurisdictions was actively pushed, and the number of decisions they rendered in the district grew from 177 in 1927 to 476 in 1928. In Vermeulen's view, of course, native courts were to be organized along the same lines as the projected native councils. However, to most local chiefs who had been handling customary litigation on a de facto basis for years—and who had been collecting traditional court fees in the process—the new system was largely a nuisance, and this was reflected in the attitudes of the people. As the district commissioner himself admitted,

> natives show little alacrity in bringing their palavers (i.e., their civil suits) to the new courts. In their minds, this is another one of the white man's contraptions. But what gives them pause in most cases . . . is the fear of bothering a chief who prefers the old system.[87]

Vermeulen's own description of the proceedings in these new courts, however, helps explain why the Africans did not regard the new courts as an expression of their judicial autonomy:

> The Administrateur, who is entitled to preside the court [article 7],

never misses an opportunity of bringing to the attention of native judges certain abusive aspects of customary law of which we disapprove. Whether or not he actually presides over a session, he has occasion publicly to exhort black judges to show equity, to defend the weak, the oppressed, and the duped.[88]

The "official" native courts organized by the 1926 decree did remain a part of the colonial system to its end, but their importance as a facet of Vermeulen's proposed reorganization of the native administration system of Lunda declined rapidly. At Kapanga itself, the number of cases transacted fell from 128 in 1928 to 41 the following year.

The concept of native treasuries did not fare much better. These began to function in 1928 in Lulua, and their total receipts for that year amounted to 31,336.20 francs (approximately $3,000 today) for the entire district, of which the portion derived from court fees and fines (25 percent) barely sufficed to cover court expenditures. In fact, the bulk of the receipts came from recruitment premiums previously rebated to the chiefs by labor recruiting agencies, notably the BTK. With Lunda participating only marginally in the Katanga employment boom of the 1920s, such receipts were bound to remain limited, and, when the depression all but ended the recruiting drive,[89] that particular source of income fell to almost nil.

Another of Vermeulen's pet projects was the development of so-called model villages, a program involving the rehabilitation of selected villages (usually the chief's residence) by the utilization of local resources, with such features as the encouragement of the use of adobe brick constructions, the training of local craftsmen, the development of simple sanitation measures, etc.[90] The program met with some degree of success, particularly at Kaumba's Musuumb and in the Kapanga area, and the district commissioner was praised for his dynamic action; but his political views continued to be ignored.

Nor were the provincial authorities greatly impressed with the latest evidence of the Mwaant Yaav's far-reaching prestige, namely his intervention at the request of a Belgian official in the native administration problems of Kwango. In the spring of 1928 an *administrateur territorial* from the District of Kwango solicited the Mwaant Yaav's intervention to arbitrate the conflicting claims of local Lunda headmen. Kaumba was naturally delighted to oblige; a mission consisting of some ten courtiers led by the *Mwanaute* Kankuluba Naweji left for Kwango on June 16, 1928, and, traveling by way of northern Angola,

Tshikapa, and Feshi, it reached Kasongo Lunda, capital of the *Kiamfu* of the Bayaka in May of the following year. Judging from the reaction of the Belgian administrator at Kasongo Lunda, who requested that no more delegations of this sort be sent in the future,[91] it would seem that the mission was rather counterproductive in Belgian eyes; but from the Mwaant Yaav's viewpoint, it served to reestablish long discontinued relations with western Lunda communities.[92] This was not the first instance of the Mwaant Yaav's continued prestige beyond the boundaries of the district: in the spring of 1927 Chief Kanongesha from Northern Rhodesia and Chief Mwazaze from Angola had traveled to the Musuumb to pay homage to the Lunda paramount.[93]

Rather than stressing such doubled-edged arguments, however, Vermeulen preferred to underline the poor performance of Cokwe chiefs and the historical affinities between Lunda and Cokwe. In October 1930, shortly before going on leave, he restated once more the case for a Lunda-Cokwe secteur:

> We must admit that the experiment we made by organizing fully independent Tshokwe chefferies has produced few practical results thus far. . . . Let us not forget that all these [Cokwe] chiefs cannot be regarded as traditional chiefs, but as representatives of some ruling families from Angola placed at the head of disparate groups lumped together into chefferies. Nor should we forget that outside these seemingly homogeneous groups, many Tshokwe clans have settled in the midst of Lunda sous-chefferies and have willingly shared in their political institutions for a long time.[94]
>
> Furthermore, he added, the dissensions that seem to have existed these past few years between the Tshokwe and the Lunda are wholly illusory. They were artificially created by the currently recognized chiefs who used them as a springboard for their policy and as a source of prestige.[95]

None of these considerations were new, of course — which is not to say that they were well-founded — but to the provincial authorities the resort to the concept of secteur was viewed as a mere "device that would subsequently allow the political integration of Tshiokwe and Aluena communities into the Lunda chefferie under Mwata-Yamvo," a development which they regarded with good reason as "totally opposed to previous instructions."[96] The vice-governor stood on firm ground when he noted that according to the governor general's explicit directives, secteurs should be resorted to only when dealing with "a galaxy of small, independent chefferies having no ethnic links between them" and "should under no circumstances be used as a means of reconstructing large native states."[97] It was all right to hold occasional meetings of subchiefs in each territoire, even perhaps an annual meeting for the entire district, as long as only Lunda subchiefs were involved and such councils were not institutionalized. As for the historical and cultural arguments advanced by Vermeulen, they were swept aside rather than refuted.

> It is politically useless to reopen the controversy concerning the ethnic affinities and historical links between the Lunda, Tutshiokwe and Aluena. It may be that Lunda is the cradle of the Tutshiokwe and of the Aluena, but a long cycle of migrations has in any case created such a breach as to modify completely the original ethnic characters that are presumed to be held in common. It seems that no traditional authority was ever exercised by Mwata Yamyo over the Tutshiokwe and the Aluena. All we recognized were the land rights of the Mwata Yamvo over the land areas occupied by heterogeneous clans — and such rights may even be debatable in the case of the Tutshiokwe who, according to all evidence, seized [their lands] through a conquest beginning around 1880 and ending only under pressure from European forces. It is appropriate to maintain the complete independence and political freedom of the Tshiokwe and Aluena groups.[98]

It matters little that Vermeulen had gone on leave a few days before this reply was written; the repudiation of his plans for Lunda was categorical and, as it turned out, final. Over the previous decade District Commissioners Labrique, Van den Byvang, and Vermeulen (not to mention Duysters) had successively come to recommend a policy which, if applied, would have involved a return to the system inaugurated in 1915 by District Commissioner Gosme and some sort of restructuring of the Mwaant Yaav's authority over the whole of Lunda.

None of these projects, however, could escape the criticism of attempting to revive a system whose traditional mechanics were imperfectly understood and which, in any case, had ceased to function normally for at least two generations. Nor could they, unless Belgian officials were prepared to abstain from dealing directly with the subchiefs, prevent this latter group from seeking to establish a functional relationship with local administrators, thus short-circuiting the hierarchical channels which supposedly connected them with the Mwaant Yaav. Vermeulen's resort to the concept of secteur, evidently meant to apply to segmented or fragmented societies, confused the issues rather than clarifying them in the case of Lunda. Despite the overall validity of many of his arguments regarding the historical and cultural unity of Lunda, his insistence on creating a hierarchy of nontraditional councils and linking them with the controversial system of native courts and native treasuries left him open to the fundamental objection formulated against the secteurs by Monsignor de Hemptinne: "Native common sense . . . will never recognize our capacity to *create* a *customary* institution: this is a contradiction in terms."[99]

In the final analysis, however, the true reasons of the opposition to the preservation of large native states must be sought in the underlying premises of an imperial philosophy that was fundamentally assimilationist and interventionist. A man like Vermeulen, with his "model villages," his preoccupation with introducing new cash crops, his pleas in favor of giving traditional chiefs a Western education, was probably as anxious to change African society as de Hemptinne, who had described Belgium's policy as follows:

> The civilizing nation means to bring the black race, cautiously, slowly and surely, to a more exalted conceptualization of human existence; it believes in the perfectibility of natives and invites them to take part in its endeavors, its labors, its civilization, and in the general progress of humanity . . . . As a transition and in order not to skip the stages of a normal evolutionary process, the structures of traditional systems will be maintained, but with the determined intention of causing this traditional authority to evolve in our direction and of eliminating it wherever it might become ossified or stand in the way of the civilizing process.[100]

The provincial authorities themselves left no doubt that any thought of restoring the Lunda empire under Belgian auspices should be ruled out. Describing the situation in the Lulua district in their annual report for 1930, they concluded:

> Political organization, which is very delicate because of the interpenetration of the Lunda and Tshiokwe elements, is not perfectly adjusted yet. A past tendency to give predominance to the former group resulted in further complicating the situation. We must return more strictly to the policy decided some years ago by the pro-

vincial authorities, which is based on the principle of respecting the independence of each of the ethnic groups involved.[101]

With this cavalier paragraph, a chapter of the history of Lunda closes. Never again will the possible revival of the Lunda empire be discussed or suggested officially under Belgian rule.[102] Already the depression is darkening the horizon: within a few months, Belgian colonization will enter one of its most dramatic crises, one result of which will be an administrative reorganization of such magnitude that the whole system will be shaken to its foundations. During the crisis years, traditional polities will be further eroded and forced to fit even more narrowly into patterns dictated by bureaucratic and economic convenience.

Despite the setbacks and frustrations experienced under Belgian rule, Lunda had managed to preserve and even rebuild some of its damaged structures. The policy of ethnic autonomy introduced after the war had opened the way for a subversion of traditional relationships and ushered in the threat of dismemberment, yet Belgian authorities themselves had arested the process and upheld the Mwaant Yaav's paramountcy, at least on paper: in 1930, only six of the twenty-nine chiefs of the Lunda heartland were not subordinated to Kaumba.[103] Whether they realized it or not, Lunda traditionalists had reason to be satisfied with their position at the opening of the decade. Within a few years, however, conditions would be much bleaker, and Lunda would drift even more irrevocably away from its past glories.

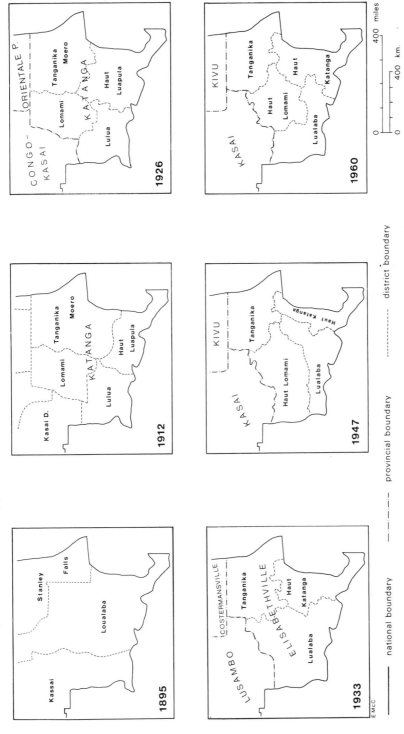

Administrative Divisions of Katanga, 1895-1960

5

Stagnation and Change:
The Depression and the Crisis
of Colonial Administration

Throughout the Congo the depression ushered in a general reordering of priorities and compounded the disruptive effects of colonial rule. At the time of Belgium's takeover of the Congo in 1908, an economic system based upon the collection of natural products and requiring only a minimal infrastructure was still largely in place; in that year rubber and ivory still accounted for 84.5 percent of the Congo's exports. But Leopoldian policy soon gave way to a more systematic form of exploitation involving the development of plantations and mines. Salaried manpower grew from 47,000 in 1917 to 427,000 in 1927, and villages were drained of their adult male population to such an extent that alarmed administration and missionary circles finally imposed quotas on labor recruitment. In Katanga, the Union Minière, which had begun production in 1911 with a modest output of 998 metric tons of copper, was turning out 57,886 tons by 1923 and 137,000 tons by 1929.[1] Three years later, however, copper production had relapsed to 54,000 tons, and the number of salaried workers had shrunk to 291,961, a drop of nearly one-third from its 1929 level. Colonial budgets were slashed mercilessly, and by a decree of June 23, 1933, the number of districts was reduced from 23 to 15 while the number of territoires was cut, even more drastically from 205 to 105, thus expanding their average size to nearly 9,000 square miles. Native administration followed suit shortly thereafter: a decree of December 5, 1933, gave legal status to the secteurs and provided the necessary instrument to streamline the rather haphazard patchwork that had evolved over the previous fifty years. Even after a decade of informal amalgamation of scattered populations into secteurs there were still almost 5,000 native administration units in 1930, but by 1938 that number had been reduced to 1,552, of which 340 were secteurs.[2]

Even though the indiscriminate recognition of "chiefs" which had taken place until the end of the first world war cannot by any means be attributed exclusively, or even predominantly, to a respect for traditional institutions, there is no doubt, on the other hand, that the 1933 decree on native administration and the ensuing consolidation of smaller units into larger ones was prompted primarily by reasons of economy and not by any real desire to establish the foundations of a viable system of indirect rule. Even so, the inevitable decline of direct administrative supervision caused by the drastic cuts in the ranks of European officials, combined with the extension in the size of the average area administered by an African chief might have given the latter a greater amount of de facto autonomy in the handling of local issues. In fact, what the 1933 reform of native administration really meant was that, more than ever before, African chiefs would be turned into auxiliaries of the depleted civil service. "While respecting traditional organization, the legislator wanted to establish a single administrative system: he made the chieftaincy the lowest echelon of the administrative organization and the chief a functionary integrated into the system without prejudice to his traditional role."[3]

The overall degree of governmental interventionism was not substantially affected by the depression. Indeed, the need to police the tens of thousands of Africans who were being thrown back on subsistence agriculture, and at the same time to avoid the excessive demobilization of a labor force which had taken such coercion to be raised in the first place, implied a host of new controls. Chiefs were saddled with new duties which, under more favorable economic circumstances would have been assumed by European officials. Among these was the enforcement of the new, generalized measures regarding the compulsory cultivation of certain crops. Introduced on a limited scale during the first world war, the obligation for African communities to plant certain designated crops had originally been intended to secure a stable food supply for government employees, soldiers, road gangs, etc., while at the same time bringing the African peasant within the nexus of a monetary economy. The collapse of the rubber market and the general debacle of the *Raubwirtschaft* system which had characterized the first twenty-five years of colonial rule also persuaded some

Belgian circles that compulsory cultivation could induce Congolese villagers to turn to the production of export crops, just as West African peasants had taken to the growing of cocoa. In the eyes of a man like Alexandre Delcommune,[4] such a policy would have had the advantage of halting the erosion of rural communities and, in one form or another, similar views continued to be professed until the end of Belgian rule in the Congo.

Several factors, however, combined to frustrate these designs: the hostility of mining firms and other labor intensive enterprises whose recruiting efforts might have been blunted by a stabilized, prosperous peasantry; the competition of large-scale plantations controlled by European interests such as Unilever; the choice of compulsory crops, which was often dictated by reasons alien to the genuine interests of the Africans, such as the preoccupation with alleviating structural weaknesses of the Belgian economy[5] or the desire to minimize competition between African and European producers.

With few exceptions, African communities did not substantially benefit from the relative prosperity of the 1920s in the form of high prices, but, with the depression, what limited economic incentives might have existed for them to persist in the cultivation of cash crops were largely nullified, and compulsory cultivation—now to be enforced with the assistance of African chiefs—became, like the manpower stabilization programs initiated by the mining firms, a way to prevent economic demobilization by ensuring that the Congolese peasantry did not entirely drop out of the monetary economy.

Many of the above developments, however, do not apply in the case of Lunda, which entered the 1930s with an economy still reminiscent of the days of the Free State. As shown in Table 2, natural produce still accounted for half of the exports passing through Dilolo in 1931.

Remote, poorly endowed, and sparsely populated in comparison with neighboring areas, Lunda had been ignored by the Free State, neglected as a district of Congo-Kasai province and bypassed in favor of more promising regions after its transfer to Katanga in 1912. World War I brought a limited and short-lived prosperity to the farming population in the form of high food prices. Manioc flour fetched as much as one franc per kilogram at Kafakumba during those years and the proceeds from the native tax (*Impôt Indigène*) rose from 33,922 francs in 1914 to 108,646 francs in 1918, a level that would not be surpassed until ten years later. With the end of the war, however, the

Table 2. Exports through Dilolo, 1931.

Product	Value
Ivory	75,450 francs
Wax	1,080,000
Native artifacts	194,000
Skins	532,000
Lime	540,000
Cement	1,350,000
Total	3,771,450

Source: District de la Lulua, Rapport politique 1931, p. 98.

demand for food declined sharply: the price of a kilogram of flour fell to ten centimes at Kafakumba and to five centimes at Kayoyo in 1920. Bearers were paid so little that it cost more to have a ton of freight shipped by rail from Elisabethville to Bukama than to have the same load carried the approximately equal distance from Bukama to Sandoa.[6]

The decline in the volume of monetary transactions was reflected in the sharp drop of native tax proceeds (Table 3), which fell in 1921 below their 1913 level, and by the decrease in the purchase of manufactured articles. "The natives here can no longer afford textiles," the district commissioner noted. "Everybody goes around wearing skins."[7] Only the newly opened mines at Musonoi and Busanga (tin) in the neighboring district of Upper Luapula offered a limited outlet for the sale of native food crops. Manioc flour sold between 65 and 80 centimes in the mining camps, but the severe drop in world copper prices caused the mine at Musonoi to be closed down between 1922 and 1928.

In fact, the agricultural capacity of Lunda was too weak to regularly produce an exportable surplus. A country of poor soils and upland marshes, with fewer than two hundred miles of all-weather roads in 1922, the district remained susceptible to recurrent food shortages as late as 1928. Relatively small increases in the demand, created by such extraneous factors as the presence of troops, the opening of a mining camp, the building of a road or, in the late nineteen-twenties, of the railroad connection to Angola, could drive up prices almost overnight, and were met by the local peasantry to the detriment of their own food

Table 3. District of Lulua: proceeds of native tax, 1913-1931.

Year	In current francs	In gold francs
1913	46,529.50	46,529.00
1914	33,922.50	33,922.50
1916	72,000.00 (est.)	72,000.00 (est.)
1917	95,682.00	95,682.00
1918	108,646.00	108,646.00
1919	90,945.35	90,945.35
1920	243,739.40	60,934.85
1921	163,306.35	40,826.58
1922	122,496.95	30,624.24
1923	225,572.35	56,393.09
1924	475,567.30	95,113.46
1925	493,313.70	98.662.74
1926	611,536.00	97,845.76
1927	792,386.80	104,010.28
1928	886,496.40	115,244.53
1929	1,435,135.75	186,567.65
1930	1,764,504.00	229,385.52
1931	2,191,727.00	284,924.51

Source: District de la Lulua, Rapports politiques, 1913-1931.

needs. Even the area around Sandoa, where a relatively stable demand was created by the presence of district headquarters after 1915, was almost chronically short of the most basic local products during the last months of each year.

This explains why, despite its vested interest in seeing local farmers develop their monetary income, the district administration had to forbid the export of foodstuffs from Lunda to the newly opened diamond fields of Angola.[8] The supply of native produce to the African crews building the BCK railroad line across northwestern Katanga was already draining the area of its own food resources, to the detriment of the needs of the small Force Publique contingent stationed in the district. In addition to the native crop of maize, a number of European grain crops, mostly wheat and buckwheat, were tried without much success by Belgian officials. Only toward the end of the decade did the production of native food crops finally reach acceptable levels, but low productivity and high transportation costs combined to price them out of the market in the mining areas: thus the 1928 peanut and rice crops found no buyers. It was clear, District Commissioner Vermeulen observed somewhat despondently, that the commercialization of food crops only made sense on a short-distance basis, i.e., for such traditional outlets as the Force Publique and the railroad crews. "The economic activity of this District," he went on, "will remain insignificant as long as it cannot export high-grade products."[9]

As had been the case a few years earlier with the construction of the BCK railroad or the opening of mines in the western portion of the Katanga copperbelt, the building of the Tenke-Dilolo rail line connecting southeastern Katanga with the Benguela railway through the southern portion of Lunda after 1928 briefly stimulated the production of food crops in the areas adjacent to the railroad's path, but stagnation remained the rule everywhere else: only fifty tons of produce were commercialized at Kapanga during the year 1929. Even in the south, the gradual advance of the rail crews left behind a trail of depleted food reserves and no lasting prosperity.

In terms of cash crops, the situation was even less satisfactory. The collapse of native rubber prices under the competition of Asian plantation rubber destroyed one of Lunda's few exploitable resources. By the middle of 1918, rubber prices were so depressed that the natives of Kapanga territory could make more money by hiring themselves out as bearers for a ten-day journey to Sandoa and back than by collecting rubber for an entire month. The monthly output of native rubber from Kapanga dropped from one or two tons in 1916-1917 to 70-150 kilograms in the middle months of the year 1918.[10] Yet, the paucity of alternative resources kept rubber collection alive throughout Lunda for almost a decade after the rubber crisis of 1913. Years of indiscriminate exploitation had depleted the reserves of native rubber. Entire villages from northern Lunda had to travel for weeks into Kasai in search of untapped supplies and the prices paid by the Compagnie du Kasai, which enjoyed a virtual monopoly, were stigmatized by the Belgian administrator for Kapanga as "scandalous."[11] Still, in 1921, the district managed to export 123.5 tons of rubber, but four years later exports had dwindled to 2 tons and they remained negligible until the second world war.

With the decline of rubber, wax (an old staple of the caravan trade) became by default the chief export of the district, albeit a rather precarious one. A traditional activity with the Cokwe, wax collection was regarded by the Aruund as a demeaning occupation and never took hold in the north. From 86.5 tons in 1921, wax exports from the district had declined to 13.2 tons in 1931. Unfortunately, no alternative export crops had been developed in the meantime to take the place of rubber and wax. Hundreds of thousands of palm kernels were distributed to the villagers in the hope of fostering the development of native plantations, but Belgian officials were soon forced to admit that palm trees simply did not grow south of Sandoa. Even in the north of Lunda results were mediocre: only 300 kilograms of palm oil were produced in the Kapanga area in 1928.

Year after year the colonial administration tried to introduce new cash crops — kapok trees, flax, Fourcroya (a variety of sisal), coffee, guavas, oranges — only to meet with the same discouraging results. Even if we allow for the hortatory style of official reports throughout the world, there is something both pathetic and comical in District Commissioner Labrique's description of the sale of 400 pounds of Fourcroya rope in 1921 as the appearance of a promising new export.[12] Even when the procurements branch of the provincial government agreed to order their entire office supply of twine from the District of Lulua, sales remained ridiculously small: 450 pounds of rope coming from a model plantation at district headquarters were sold to the post office in 1923, after which the whole operation was simply abandoned. As for cotton, it was not successfully introduced until the late nineteen-thirties.

Cattle breeding, another potential resource, also turned out to be a disappointment. A few head of beef cattle had reportedly been introduced at the Mwaant Yaav's court by early Portuguese traders but were not successfully bred. The Cokwe themselves brought some head of cattle after settling in Lunda. According to one source, the Cokwe lost a portion of these cattle in the form of tribute to the Mwaant Yaav when Belgian officials attempted to restore the latter's authority over the whole of Lunda, and they liquidated the rest for fear of having it confiscated by the Europeans, so that only fifty head remained by 1915.[13] While this may be an exaggeration, the fact remains that only a few years later the total number of beef cattle in the district was officially put at 528, nearly half of them owned by non-Africans (Table 4). Many of these cattle were diseased and of low yield. At no point did

Table 4. District of Lulua: beef cattle, 1923-1925.

	1923	1925
Owned by Africans	287	339
Owned by non-Africans	241	289
Total	528	628

Source: District de la Lulua, Rapport politique 1923, p. 33; ibid. 1925, p. 31.

they represent a potential source of cash income to the Africans, even though the supply of meat to the mines and railroad camps would have provided an adequate outlet for the production of animal proteins. Indeed, the supply gap was eventually filled, at least in part, by the European corporations themselves through subsidiaries. In 1931 the number of cattle in the southern part of the district was estimated at less than 2,000 head of which 1,465 were owned by a single cattle-breeding concern, Agricomin, which was based at Mutshatsha, the railroad maintenance center for the entire area. The rest were more or less evenly divided between Africans and non-Africans (mostly missionaries) but were destined to prompt extinction by order of the veterinary services to halt the spread of an epizootic disease.[14] The situation was hardly more promising with respect to small cattle, as shown by Table 5.

Lunda's remoteness from the industrial and administrative centers of the Congo must account, at least in part, for its economic stagnation, but the lack of adequate transportation was only one facet of the problem. At the end of the war there was not a single mile of all-weather road (let alone of rail) in the entire district, but, beginning in 1920, credits were made available for a fairly extensive road-building program. In the marshlands of central Lunda the task was often monumental: several sections had to be built on levees, and countless small rivers had to be bridged. The roads, which had initially been planned to be 3.5 to 4 meters in width, soon had to be widened to 6 meters in anticipation of automobile traffic, which first appeared in 1922, and by 1925 Sandoa could be reached by truck from Bukama, on the BCK rail line.

The strain on local manpower was enormous. In 1920 alone, 73,584

Table 5. District of Lulua: Distribution of small cattle, 1923.

Territoire	Goats	Sheep	Pigs
Kapanga	874	285	124
Sandoa	344[a]	159[a]	52[a]
Dilolo	n	n	n
Kafakumba	153[a]	57[a]	24[a]
Kayoyo	820[a]	200[a]	191[a]
Kinda	1,500	20	275
Total	3,691	721	466

Source: District de la Lulua, Rapport politique 1923, p. 33.

[a]Female animals only.

n = negligible

days were contributed by requisitioned laborers (against 11,483 by prison gangs) on public works projects in the district, a figure which does not take into account the many instances — duly noted by Belgian officials — where chiefs themselves "volunteered" the services of their entire village for the construction of roads, levees, or bridges.[15] The administrator stationed at Kayoyo warned in 1923 that the extent of road construction during the two previous years had prevented the villagers from planting their crops and that serious food shortages would ensue in the area under his jurisdiction.[16]

The gradual appearance of trucks in Lunda after 1925 did not significantly alter the economic situation. For one thing, of course, there was little produce to transport; for another, trucking costs were prohibitively high for most producers or small traders. As had been the case earlier with the BCK railroad, trucking turned out to be more expensive than porters. In 1926, however, the porterage system was abolished along all roads accessible to trucks and, while this measure liberated the Africans from a particularly odious form of disguised servitude, it gave the trucking firms a quasimonopoly and made local producers even less competitive while adding to the cost of imported articles. Official hopes for an escape from economic stagnation now turned increasingly to the advance of the Benguela railway, which had been anticipated since the beginning of the century (the Benguela Railway Company was founded in 1903), and more specifically to the construction of the Tenke-Dilolo extension, which had been under study for over a decade and would connect the Katanga copperbelt with Lobito across the southern part of the district.

An almost chiliastic aura surrounded the opening of the railway: the modern world was finally coming to the district, and prosperity would follow in its wake. Work began on the railroad in 1928, and the line was inaugurated on July 1, 1931, but it soon became clear that it could not be the hoped-for panacea. The railway closely followed the Congo-Zambezi watershed and thus remained well to the south of the historical center of Lunda. The Kapanga region was actually closer to the BCK line than to the new Tenke-Dilolo section, and the cost of transporting crops to the nearest railroad station — assuming that there would be crops to export — remained a serious handicap for the economy of the region.

In fact, rather than promoting the development of the whole district, the opening of the railway polarized what little economic activity there was and further contributed to shifting the economic center of gravity of Lunda away from its traditional political center, a process which had actually begun some fifty years earlier. In 1930, twenty of the thirty-one business firms operating in the district were based in the three southern territoires (Dilolo, Luashi, and Kayoyo-Mutshatsha), but in 1931, the year the rail line was completed, the number of firms increased to forty-six, of which thirty-one (i.e., as many as had existed in the entire district the year before) were located on or near the rail.[17] In any case, the onset of the depression made the opening of the railway seem somewhat anticlimactic, and the vice-governor, commenting on the economic situation in the district, was forced to recognize that

> because of the overall depression, the District did not receive from the advent of the railway the impulsion which might have been anticipated; a certain degree of activity could be maintained, however, thanks to the work needed by the completion of the track. As far as food crops are concerned, the situation is unfortunately such that we should no longer encourage production, except to the extent that the natives can sell their produce.[18]

The lack of economic development of Lunda had two other important consequences: the sparseness of European settlement and a comparatively low rate of socio-economic change, both of which continued to be apparent until the end of the colonial period, and indeed to the present. The total number of Europeans residing in the District of Lulua in 1920-1921 was only sixty-four. Belgian officials accounted for approximately one-fourth of that number, and missionaries were still represented only by a small group of American Methodists. In 1913 the Methodist Episcopal Church had founded the first Christian outpost in the Lunda heartland. Their station, which was named "Lusambo," was located some four miles north of Sandoa and was staffed by five American nationals and a handful of African catechists. By the end of the first world war, the missionaries were offering rudimentary instruction in Kiruund at two schools, one at Lusambo, the other at Kapanga. Another station was opened in 1920 at Kinda in Luba territory, while African catechists were operating in the chiefly villages of Muteba, Mwene Matanga, Mbako, and Kayembe Mukulu. Catholic missionaries did not appear in Lunda until 1922, when a first station was founded at Sandoa by the Franciscan order.

Small as their number was, traders thus represented the majority of the non-African population during the early nineteen-twenties. There were twenty-five trading posts in the district in 1920, all of them operated by Portuguese nationals, except for those run by the Compagnie du Kasai, which ended its activities in the Lulua during 1921. As a result, the number of trading establishments dropped to nineteen in 1921, then climbed to twenty-eight at the end of 1922, only to relapse to twenty-five a year later. Most of these posts were located in the southern part of the district; a single European firm covered the Sandoa and Kapanga regions and, in 1928, there was only one small trading post at Kapanga and one at Kafakumba, once the administrative headquarters of the district. The situation had not improved much by the time the effects of the depression began to be felt in the Congo: there were forty-nine trading posts in 1930 and fifty-five the following year, but much of this modest growth had been artificially generated by the building of the railroad and did not survive its completion. Relatively few Belgians were attracted by the mediocre commercial prospects of Lunda; of the fifty-five trading establishments active in 1931, only ten were in Belgian hands and the rest were held by Portuguese nationals, South Africans, and British subjects of Levantine extraction.[19] The prospects for European agricultural settlement were even less satisfactory. There were only two settlers in the district in 1920, both of them Portuguese, and five the following year, but District Commissioner Vermeulen felt in 1928 that, because of low population density and lack of available manpower, Lulua could not possibly accommodate more than ten economically viable European farms.[20]

Labor migrations and salaried employment usually rank high on the list of reliable indices of socio-economic change, if only because they are readily quantifiable. The opening of the mines of upper Katanga created almost overnight a tremendous demand for African labor which the sparsely populated copperbelt could not conceivably meet. The existence of a rail link between Elisabethville and the Rhodesias made it possible, from the very beginning, to recruit laborers in southern Africa, and this pattern was only gradually compensated by the recruitment drives carried out in Luba territory along the BCK rail line. As late as 1927 non-Congolese employees still outnumbered both Katangese and "other Congolese" on the payrolls of the Union Minière, and at the end of 1930 there were 11,015 alien Africans among the 34,835 black residents of Elisabethville.[21]

Beginning in 1919 the major employers of upper Katanga organized a sort of cooperative patterned after the Witwatersrand Native Labour Association and the Rhodesian Native Labour Bureau for the purpose of centralizing labor recruitment operations. The Bourse du Travail au Katanga (BTK, superseded after 1927 by the Office Central du Travail au Katanga, OCTK) carried out its recruiting activities almost exclusively in Katanga, and while it did not entirely eradicate private recruitment,[22] it nevertheless represented by far the most important catalyst for turning the Katanga peasantry into an industrial proletariat. It is all the more significant, therefore, to observe that Lulua was the only one of the four districts in Katanga where the BTK failed to carry on its recruiting activities during the first ten years of its existence. It was not until after World War I that the first laborers were recruited from this area, which, although it accounted for approximately one-sixth of the total population of Katanga, supplied only 12.2 percent of the manpower recruited in that province by the BTK/OCTK from 1913 to 1931 (see Table 6).

These laborers were overwhelmingly absorbed by the mining and railroad companies; together, their share of the labor force recruited by the BTK/OCTK between 1914 and 1930 ranged from a minimum of 77.5 percent (in 1914-1915) to a maximum of 98 percent (in 1918-1919).[23] For the people of Lunda this meant, in most cases, being

Table 6. Labor recruitment[a] in Lulua and Katanga by the
BTK/OCTK, 1910-1931.

Year	From District of Lulua				From all Katanga	
	Men	%	Days	%	Men	Days
(1910-13)	—	0.0	—	0.0	(9,000[b])	(not available)
1913-14	—	0.0	—	0.0	3,584	691,100
1914-15	—	0.0	—	0.0	7,549	579,850
1915-16	—	0.0	—	0.0	5,258	737,325
1916-17	—	0.0	—	0.0	5,348	1,244,700
1917-18	—	0.0	—	0.0	4,653	1,112,275
1918-19	—	0.0	—	0.0	4,060	917,825
1919-20	11	0.2	2,550	0.2	7,150	1,660,075
1920-21	1,058	14.2	321,125	12.5	5,227	1,579,975
1921-22	1,512	18.0	432,375	15.0	6,340	2,029,075
1922-23	1,135	13.2	394,150	12.5	6,535	2,305,225
1923-24	1,428	17.1	514,075	17.1	7,184	2,496,875
1924-25	2,223	26.1	796,925	26.4	8,499	3,019,450
1925-26	828	13.2	285,575	10.9	5,714	2,427,125
1926-27	355	4.8	121,325	4.2	7,014	2,876,450
1928	1,524	24.0	812,550	27.2	6,345	2,984,523
1929	1,558	22.6	909,600	21.8	7,014	4,164,520
1930	782	17.2	460,170	14.2	4,555	3,246,782
1931	88	21.2	38,310	15.5	416	246,771
Total (1913-31)	12,530	12.2	5,088,790	14.8	102,445	34,320,971

Source: Rapport annuel de l'Office Central du Travail au Katanga, 1931 (Brussels, 10 March 1932); Rapport au Comité Local de la Bourse du Travail du Katanga, 1920 (Elisabethville, 1920).

[a]Figures exclude laborers recruited outside Katanga as well as contract renewals by laborers recruited in Katanga, but include laborers recruited directly by the Union Minière in Lomami after 1926 and those recruited directly by the Géomines in the Tanganika-Moëro after 1929.

[b]Estimated (not included in total).

transplanted anywhere between two hundred and five hundred miles from their homes for long periods of time. The average term of the contracts signed by the African recruits of the BTK/OCTK in 1920-1921, was slightly more than one year and had increased by 1930 to nearly two years.[24] Of course many of these recruits signed up for another term upon completing their original contract, or found another form of employment in the industrial regions. One way or another, these adult males were lost for their village communities, for, while many of them might send remittances to their kinsmen and remain loyal to their clan and tribe, their physical absence from the village had direct and obvious repercussions upon the agricultural and demographic productivity of rural communities.

It has already been noted that Lunda not only failed to develop any significant cash crops during the interwar period but was also subject to local but recurrent food shortages during the 1920s. Detailed statistics regarding demographic trends are not available for this period, but at the turn of the decade the district of Lulua lagged far behind every other region of Katanga in terms of natural population increase (Table 7) and was actually on the verge of serious depopulation — a trend which became even more obvious in the central and northern portions of Lunda in subsequent years.

Table 7. Katanga: Birth and death rates in tribal areas, by district, 1930-1931.

District	Birth rate (per thousand)		Death rate (per thousand)		Growth rate (per thousand)	
	1930	1931	1930	1931	1930	1931
Elisabethville	38.51	54.45	20.34	25.91	18.17	28.54
Haut-Luapula	35.42	32.99	26.40	21.67	9.02	11.32
Tanganika/Moëro	38.96	35.78	30.24	23.87	8.72	11.91
Lomami	36.58	28.74	27.65	25.21	8.93	3.53
Lulua	33.01	27.90	34.43	25.95	-1.42	1.95
Average rates	36.25	33.86	28.14	23.20	8.11	10.66

Source: Vice-Gouvernement Général du Katanga, Service des AIMO, Rapports annuels 1930 and 1931.

It does not seem plausible to ascribe the demographic stagnation of Lunda solely, or even primarily, to the effect of labor recruitment; after all, the other districts contributed a proportionately larger share than Lulua to the constitution of an industrial labor force without suffering the same demographic loss of substance. To be sure, recruitment by the BTK/OCTK represents only a part (albeit the most important one) of the rural manpower drain between 1910 and 1930, but there is no reason to believe that Lunda contributed more than other regions of Katanga to the labor migration flow in the form of voluntary enlistments—quite the contrary. But two other factors are worth mentioning. First, intensive labor recruitment in Lunda was concentrated over a relatively short period: the district of Lulua supplied 11,649 valid men—i.e., ninety percent of its total contribution to the recruitments effected by the BTK/OCTK—during the nine years from 1920-1921 to 1929, while in the other districts the recruitment drive was spaced over a period of some twenty years. Secondly, we know that a notoriously high percentage of recruits from Lunda had to be turned away for failing to meet the physical requirements of the labor recruiters.[25] It is not inconceivable that the combination of poor nutrition and of generations of reverse selection in what had long been a center of human diffusion in south central Africa had left the population of Lunda physically diminished and thus more vulnerable to the loss of its ablest males.

Census data collected by the Belgian administration during the 1920s (Table 8) offer little assistance. From 1920 to 1925 the figures mentioned in the table are those of populations actually covered by census, and their growth over the years tends to reflect the extension of administrative control rather than actual demographic increases. Official sources estimate the actual population at 20-25 percent above the census figures, but this estimation is in all probability far too low, at least for the first few years. After 1925 a new, more ambitious census technique (*recensement par fiches*) was gradually introduced, but, until it became fully operative, the only satisfactory information regarding population is based on the yearly estimates submitted by the administrateurs de territoire. Not all these data are equally reliable, however. In the case of Kapanga, for instance, the estimates offered by Duysters during his incumbency (1926-1927) would appear to reflect his overall inclination to maximize the importance of the Mwaant Yaav's chefferie, but the tendency to overestimation seems almost unan-

Table 8. District of Lulua: population, [a] by territoire, 1920-1932.

Year	Kapanga	Sandoa	Dilolo	Luashi	Kayoyo/ Mutshatsha	Kafa- kumba	Kinda	Total
1920	20,507	22,293	10,647		10,586	13,000	13,359	90,392
1921	21,941	22,347	11,336		13,214	15,189	13,642	97,669
1922	27,724	22,347	14,241		14,816	18,936	13,540	111,604
1923	27,156	30,269	22,122		14,816	18,947	13,555	126,865
1924	27,996	39,331	27,940	8,812	11,074	18,942	13,204	147,299
1925	29,171	37,119	34,066	14,853	10,315	19,659	13,204	159,387
1926	46,509	41,723	59,350	22,015	8,048	17,637	17,214	212,496
1927	45,876	41,613	51,020	18,985	12,328	37,586	14,077	221,485
1928	37,939	43,288	42,635	20,888	13,333	30,897	14,137	203,117
1929	37,939	36,453	43,451	18,188	13,325	30,692	16,176	196,224
1930	33,618	38,173	43,699	22,215	12,197	31,435	17,934	199,271
1931/2	35,122	33,454	36,268	25,059	15,283	30,963	17,680	193,829

Source: District de la Lulua; Rapports politiques 1920-1932.
a1920-1925: population covered by census. Actual population (est.): 20-25% above census figures. 1926-1932: estimates of total population.

imous during these two years, and downward revisions of estimates during the late twenties do not necessarily reflect actual decreases in population. Wherever population estimates for 1931-1932 stand below the figures for 1925 (the last year when the old census methods were universally applied), as in the case of Sandoa, it seems legitimate, however, to speak of depopulation.

The southward shift of population mentioned earlier in this chapter is also easily discernible: the three southern territoires of Dilolo, Luashi, and Kayoyo (later Mutshatsha) together accounted for less than one-third of the population of the district in 1924, but by the end of 1931, they represented nearly 40 percent of this same population. An eastward shift is also apparent during this same period: the relative share of the total population in the district living in the three western-most territoires (Kapanga, Sandoa, and Dilolo) dropped from approximately 65 percent in 1924 to 54 percent in 1930-1932. In absolute numbers, the total population of these same three territoires went from 100,354 in 1925 to 104,844 at the end of 1931, an increase which amounts to a mere 4.5 percent over six years and can undoubtedly be described at best as stagnation.

As with the slave trade, there is no satisfactory way of evaluating the loss in human substance and dynamism suffered by a given area as a result of long-distance migration. Not all laborers recruited for the mines and railroad camps of industrial Katanga remained there per-manently, of course, The BTK/OCTK, for its part, estimated that, while it channeled 98,901 adult males into the company towns be-tween 1914 and 1933, it repatriated 51,134 to their home areas over the same period.[26] But one can hardly expect a man who has been transplanted to a remote industrial area simply to resume his place in the village after his contract has come to an end, as if nothing had happened. There is almost no sense in which the cold arithmetic of labor repatriation can be construed as somehow compensating for the loss initially suffered by traditional communities.

Repatriations, in any case, followed a pattern dictated by economic rather than humanitarian considerations. They reached a peak after the Union Minière had introduced its famous manpower stabilization program in 1928 and when the depression hit Katanga, the OCTK virtually gave up recruitment to devote itself almost exclusively to the repatriation of the masses of laborers who had been suddenly laid off by the mining companies and whose presence in the urban areas was viewed by the authorities as a possible source of disorder. In the case of the Lulua district, repatriations between 1927 and 1930 totaled 1,871 men, while the number of recruits for the same period amounted to 4,247. Between 1931 and 1933, however, recruitment dropped to a mere 98, while repatriations increased to 1,904.[27] But, as Strythagen, the director of the OCTK admitted, repatriation also raised a number of problems. "Most of the workers who have been laid off and have returned to the village accept traditional authority only with the greatest difficulty. It will take a certain time before they are readjusted to the village life."[28] With unconscious cynicism Strythagen suggested a number of possible solutions to the plight of the unemployed Africans:

> It seems that a solution to be recommended would be . . . for the government to undertake a massive program of public works . . . ; it is in the public interest to undertake at low cost projects that would have to be undertaken anyway sooner or later . . . . Roadwork could, in our opinion, absorb the unemployed provided that appropriate legislation be passed to force them to find employment. Current legislation makes it almost impossible to repress vagrancy. . . . It is truly unfortunate that because of legal technicalities we are led to encourage laziness and social disorder. . . . The compulsory registration of the unemployed should be introduced. . . . If within three weeks a man had found no employment . . . he would not be entitled to refuse any work offered to him, even outside his specialty and even for a lower wage than that indicated on his unemployment card. Refusal to comply would be penally sanctioned and, upon completing his prison term, the unemployed would be given a choice between working for a European firm or being placed at the government's disposal for at least six months. We also feel that . . . it would be appropriate to lower the cost of native manpower. The cutback of rations, which has already been decided, is a step in that direction. We must . . . roll back wages and beat back the demands of the Blacks, which have become excessive. . . . It is a matter of survival for many firms to seize the opportunity of this depression to restore certain values and to straighten up matters vigorously. . . . Finally, we feel that the native tax rate should be raised.[29]

For the natives of Lunda who had left their homes to find employment in the industrial areas, unfortunately, there were no real alternatives to sticking it out in the copperbelt or going back to the village: the lack of economic development in the District of Lulua meant that opportunities for local salaried employment were virtually nonexistent. Apart from the Tenke-Dilolo railroad, whose manpower needs declined sharply after its completion in 1931, just as unemployment in the mining areas was beginning to mount, there were no large employers in the district: the railroad and its subcontractors accounted for 70 percent of the 4,237 salaried jobs held by Africans in 1931.[30]

The relatively low degree of socio-economic change in Lunda can be measured when we know that, at the end of 1931, only 6,687 of the 193,928 African residents of the District of Lulua were living outside their traditional communities. These "detribalized" Africans represented a mere 3.4 percent of the population of the district, as against 7.3 percent within the entire province of Katanga, while in Elisabethville and its immediate hinterland (Upper Luapula District) as many as 30.1 percent of the Africans were living in a nontraditional environment (Table 9).

Table 9. "Detribalized" population in Lulua and Katanga, 1931.

	District of Lulua	Katanga Province	
No. living in non-traditional communities	(A) 6,687	(C) 87,440	A:C = 7.6%
Total African population	(B) 193,829	(D) 1,195,198	B:D = 16.2%
	A:B = 3.4%	C:D = 7.3%	

Source: Vice-Gouvernement Général du Katanga, Service des AIMO, Rapport Annuel 1931.

Economic development in a context of alien rule is, at best, a mixed blessing. The absence of mineral wealth and the low level of agricultural productivity which characterized Lunda had their counterpart in the relatively low weight of the colonial yoke. The long neglect of Lunda in the days of the Free State had been based, at least initially, on a tactical estimation of the relative threats posed by British and Portuguese claims over southeastern Congo—and, of course, negatively, on the absence of any known riches in the area—but after the importance of such preoccupations had declined, the lack of official interest in Lunda can only be explained in terms of its economic marginality for the colonizer. While it cannot be directly interpreted in economic terms, the sparseness of missionary (especially Catholic)

penetration until the mid-1920s also fits into a pattern of overall neglect which stands in sharp contrast with the activism and deliberateness of administration, business, and missionary circles in the more
productive areas of nearby Kasai and Kwilu—not to mention the
Katanga copperbelt.

The pressures of labor recruitment and the frustrations of land
alienation had a relatively limited impact in Lunda. In addition, the
traditional establishment had good reason to believe, as we have seen,
that the enforcement of *pax belgica* had not been entirely detrimental
to their survival. The combination of these elements undoubtedly goes
a long way toward explaining the virtual absence of what might be
called "secondary" resistance movement (i.e., resistance to the introduction of systematic administrative and economic controls, as opposed to resistance to the initial establishment of alien rule). In contrast with the Kwilu, where the Unilever plantations offered local
employment opportunities to the adult males, thus causing the frustrations inherent in salaried employment to be channeled directly into
the village society,[31] Lunda was introduced to socio-economic change
mostly under the indirect form of long-distance employment. The differential nature and contents of native reactions to the colonial system
offer a revealing illustration of the gap between these two regions. The
BaPende revolt of 1931 in Kwilu combined economic grievances directly related to the depression with an adventist movement leavened
by the more fundamental, yet more gradual disruption of traditional
structures.[32] In Lunda, by contrast, despite the increasingly heavy tax
burden carried by the local communities, only the latter form of
reaction can be observed.

In late 1930 or early 1931 in the Pende village of Kilamba, one
Matemu-a-Kelenia (also known as "Mundele Funji") experienced
visions in which the ancestors instructed him to kill or destroy all white
animals and objects in the land, as well as all symbols of European
rule, as a prelude to their return and to the end of white domination.
The movement spread with lightning speed and, in every village, the
people built special barns (*sombolo*) in which they placed trunks to
hold all discarded European goods: these were to be replaced by trunks
full of gifts upon the ancestors' return.[33] With its "cargo" and "ancestor return" themes, this movement is of course remarkably similar to
scores of other manifestations which have appeared at one time or
another from Indian America to the Pacific islands. The actual revolt
grew out of the adventist movement in the most straightforward

fashion imaginable: Matemu-a-Kelenia himself reportedly dealt the first blow to the Belgian administrateur de territoire, Max Balot, whose body was subsequently divided among Pende chiefs and notables.

This first incident occurred in June 1931, and by September the revolt had been ruthlessly suppressed. This was approximately the time when the first manifestations of adventist beliefs were reported from Lunda. On October 3, a Belgian sanitation agent touring Kapanga territory wrote:

> From Kalianda to Samukasa, the inhabitants have been reluctant to accept inspection; they are persuaded that in the near future all their deceased relatives shall return and bring them cattle, chickens . and food in quantity. For this reason, they have stopped farming and they slaughter all their poultry and small cattle, sacrificing them to the spirits of their ancestors who are expected to return. Why should we farm, they' say, when we will be in the midst of abundance before the rains come, and why seek medical care if we are sick? If we die, we shall return promptly with plenty of riches.[34]

Other characteristics of the Pende movements were also in evidence, notably the building of windowless ritual warehouses or "sombo" (*sombolo*).[35] Similar incidents were soon reported from Sandoa and Dilolo by local officials and by Catholic missionaries. Slight variations from the Kwilu pattern are observable, however, perhaps the most significant being the fact, confirmed by almost every witness, that *black*, rather than white, cattle and poultry were being slaughtered. Similarly, the theme of blacks and whites switching their respective skin colors was propagated in preference to the theme of the white man's demise or disappearance, which seems to have prevailed among the BaPende.[36]

Interestingly, the diffusion of the adventist rumors in the Lulua district appears to have been due almost exclusively to Cokwe illuminati (several of them adolescents) despite the fact that both Lunda and Cokwe were neighbors of the BaPende. The physical transmission of the Pende beliefs seems to have followed two major routes, both of them leading to and from the diamond-mining regions on either side of the Congo-Angola border in the Tshikapa Valley. This would explain why the appearance of the movement in the Dilolo and Sandoa regions is usually traced to Angola in the official reports, while documents from Kapanga identify south central Kasai as its proximate source. All observers concur of course in regarding the Kwilu as the epicenter of the whole convulsion.

Whether or not because of the predominantly Cokwe character of the movement, Lunda chiefs, notably the Mwaant Yaav, offered their unhesitating cooperation to the Belgian administration in its efforts to ensure its swift suppression. In the Sandoa area, the Lunda Chief Mbako, himself a future Mwaant Yaav, and the Cokwe subchief Samutoma, brother of the then deceased Chief "Mwatshisenge" Samazembe of the west central Cokwe and long regarded as a potential successor, were dispatched together by the local administrator in early January 1932 to quiet the excitement created by the adventist cult movement. Upon completing this tour of inspection, Mbako promptly denounced several village chiefs who had ordered the slaughter of black animals or the construction of "sombo" while Samutoma kept his own counsel. The reasons for his silence became apparent when the accused village chiefs charged him with having encouraged the spread of the adventist beliefs and rites. Samutoma was immediately arrested, and the whole movement was rapidly contained in the following weeks; but Belgian officials took note of the effective cooperation offered by the Lunda chiefs. The administrateur for Sandoa observed that the Lunda aristocracy "understand that it is in their interest to be on our side," while the district commissioner concluded with confidence that "the strong political organization of the Lunda will constitute a powerful barrier against subversive ideas in the north of the district, provided that the Mwata Yamvo and his court (the conservatives) are held in hand by an experienced official." In fact, however, the Mwaant Yaav's authority was not unquestioningly accepted during this episode; many villages simply refused to feed his emissaries, as was demanded by custom, and the Belgian administration had to supply them with provisions.[37] How thoroughly the adventist beliefs were dispelled is of course impossibe to determine, but it is significant that an almost identical movement reappeared in the Kasai valley in 1950.[38]

Kitawala, the earliest and largest para-Christian sect to appear in Katanga, did not manifest itself in Lunda until some years after this episode. In 1925 the Watch Tower movement had been introduced into the mining regions in a somewhat distorted form by Tom Nyarenda, alias Mwana Lesa, a native of Nyasaland who had been working on the Copperbelt. After a first wave of repression, the movement went underground, only to reappear in full force in 1930-1931, when the impact of the depression hit the copper industry. From Elisabethville it spread throughout the Katanga mining areas and, more specifically, along the rail lines, notably the CFL (Chemins de Fer des Grands Lacs) in northern Katanga, where it has remained strongly implanted. The opening of the Tenke-Dilolo rail link offered Kitawala a new road of penetration. Between May and September 1934 the area between Jadotville and Kolwezi was infiltrated by a small number of Watch Tower preachers from northern Rhodesia (less than a dozen according to one report), and thousands of Africans were baptized. Among the Kaonde of Musokantanda virtually the whole population embraced the movement which, as was frequently the case, seems to have appealed to the villagers because of its strong antiwitchcraft emphasis.[39] Belgian authorities reacted vigorously and the spread of Kitawala in southern Lunda appears to have effectively stopped for several years, despite sporadic occurrences along the rail line. As a matter of fact, colonial officials considered the area sufficiently "safe" from possible contamination to locate a village (in fact, a camp) of deported Kitawala sectators at Malonga in 1937. There were several incidents involving the "relegated" Kitawala adherents during the late 1930s, including a near mutiny in 1939, but the camp was not closed down until ten years later when it was superseded by a penal colony (COLAGREL) at nearby Kasaji.[40]

Unlike what happened among the Bemba[41] or in northern Katanga, Kitawala apparently had no serious impact on the rural areas of the Lunda heartland, and its apocalyptic attitude toward established authorities (whether colonial or traditional) had no visible effect on the chiefs of the area. In fact, after the decline of the Pende-Cokwe adventist movement of 1931-1932, popular beliefs and aspirations in Lunda seem to have been increasingly channeled toward vitalizing or immunizing cult objects (*dawa, bwanga*) rather than toward the alternative but nontraditional millennium offered by Kitawala.

By far the most widespread charm in Lunda through the late 1940s seems to have been *Ukanga,* which may have originated with the BaTetela-BaKusu of the middle Lomami valley but became enormously popular among the BaLuba and throughout Katanga.[42] Lacking the apocalyptic (and thus antiestablishment) connotations of either the Kitawala or the ancestor-return cults, *Ukanga* encountered little hostility from the chiefs, many of whom actually sponsored local versions of the charm which they could then control and utilize to re-inforce their sagging influence. Some of the Lunda chiefs who had so actively cooperated with the administration in 1932 now became involved in the diffusion of *Ukanga*. Chief Muyeye was deposed in 1934 for having favored the spread of the charm, and Chief Lumanga was dismissed in 1943 for similar reasons. More ominous, the Musuumb itself took part in the movement: Kaumba's *Nswaan Mulopw* was incriminated in 1936, and the Mwaant Yaav himself be-came implicated in the movement in 1939. He was deprived of the right to wear his paramount's medal and threatened with immediate deposition should this occurrence be repeated.[43]

Tambwe, another imported charm, ran a close second in popularity to *Ukanga* in the Lunda heartland during the 1930s and 1940s. *Tshimani,* which filtered into Lunda from the Kwango via Kasai and Angola and may have originated, according to one source, in the Uele region,[44] made its appearance in 1934 among the Ndembu of Chief Kazembe near Luashi, but did not gain such wide acceptance as the other two charms. The gradual infiltration of modernity into Lunda can be measured by the eventual superseding of *Ukanga* and other amulets based on traditional magical substances such as rainwater, human and animal parts, tree bark, and oil,[45] by new rites such as *Farmaçon* or its northern Lunda version, *Konfirmash,* which were based on the idea that the magical substance should be injected with a hypodermic needle and which came into vogue in the late forties.[46]

The administrative reorganization of 1933 was a traumatic experi-ence for the Congo, and particularly for Katanga, which not only lost the populous Kabinda district to the newly created province of Lusambo (Kasai) but also was stripped of the somewhat exorbitant status it had enjoyed since the days when its occupation had been sub-contracted to a chartered company. The old division of the Congo into four vice-governments (Congo-Kasai, Equateur, Orientale, and Ka-tanga) was replaced by the organization into six provinces, which remained virtually unchanged until the country became independent. The title of vice-governor general was replaced with the rather inglo-rious denomination of "provincial commissioner," and the name of Katanga lost its legal existence when the area was officially renamed "Province of Elisabethville" (a style followed for the five other prov-inces as well).

For the Lunda heartland, however, the changes were even more drastic. With the exception of Kinda territory, the District of Lulua had covered most of the Belgian-held areas linked to the Mwaant Yaav by tributary relationships, and even though Sandoa rather than Kapanga had been selected as the district headquarters, there had been enough of a coincidence between the district and what remained of the Lunda "empire" for successive Belgian officials over the previous twenty years to regard the preservation of the Lunda paramount's authority as essential to a successful organization of native administration in the district. Now, however, the number of districts in Katanga was reduced from four to three and the District of the Lulua was lumped with the Katanga BaLuba regions and the western Katanga copperbelt into the sprawling District of Lualaba, representing nearly one half of the province in terms of area (93,277 square miles) and population. The new district, which extended some 400 miles in all directions, was to be administered from the eccentric company town of Jadotville (which has now reverted to its pre-1928 name of Likasi). This meant that instead of being stationed 135 miles from the Mwaant Yaav's capital, the district commissioner would now be based 520 miles away by road. Where the former District of Lulua had been divided into six territoires, only two now remained. Kinda was lumped with the two large chefferies of Kasongo Nyembo and Mutombo Mukulu into the vast territoire of Kamina. The three southern divisions of Dilolo, Luashi, and Mutshatsha (formerly Kayoyo) were amalgamated into a single unit to be administered from the new railroad and prospecting center of Malonga, where manganese deposits had been identified. The northern territoires of Sandoa, Kapanga, and Kafakumba were merged into a single subdivision administered from Sandoa, now stripped of its rank as district headquarters.

Economy being the prime reason for this sweeping reorganization, the number of Belgian officials assigned to the area underwent a similar reduction. There were twenty-six administrators in the field in the District of the Lulua in 1931 (not to mention a number of desk-bound officials), but three years later only fourteen remained to cover the same area. To deal with the Mwaant Yaav's chefferie, only one low-ranking *agent territorial* remained at Kapanga, and the same situation prevailed at nearby Mato in the former Luba "empire" of Kasongo Nyembo.[47]

Under the provisions of the new decree of December 5, 1933, the only two types of native administration units entitled to legal recognition as *circonscriptions indigènes* were the chefferie and the secteur. The terms of the decree raised a number of problems regarding the organization of native administration in Lunda. Strictly speaking, there had been only one Lunda chefferie in the District of the Lulua, that of the Mwaant Yaav, and all other units (except for the six autonomous Cokwe and Lwena chieftaincies created between 1922 and 1932) were officially headed by subchiefs. But there was no place for sous-chefferies under the new system, and the administrative reorganization of Lunda became an issue almost as soon as the decree was passed.

Apart from the solution of the secteur, whose applicability to Lunda had been decisively ruled out a few years before, the only alternative was either to do away with the subchiefs, thus leaving Kaumba as the sole native authority over the entire area, or to elevate each subdivision to the rank of chefferie, with the possible corrective of merging some of the smaller units. Although the first possibility was in conformity with the overall trend toward the reduction of the number of native administration units which eventually resulted in the elimination of nearly three-quarters of them within a decade, it was never seriously considered. It would of course have meant the indirect reconstruction of the Lunda state, a perspective which Belgian authorities had always viewed with serious misgivings, if only because it disturbed the prevailing pattern of native administration in the Congo.

At this crucial juncture the Mwaant Yaav found little support even among field officials who had in the past shown the greatest receptivity to his claims. District Commissioner Vermeulen, who had returned to Lunda after his leave of absence, briefly took over the new District of the Lualaba but left it again, this time for good, in May 1934. His successor, Liesnard, had no particular interest in Lunda, and his subordinates in the field seem to have had little regard for Kaumba. The administrator for Sandoa, under whose jurisdiction the Musuumb had now been placed, complained of the passive resistance and disguised hostility he met on the part of Kaumba and his court. It apparently did not occur to the Belgian officials that the Mwaant Yaav's surliness might be caused by the fear of seeing his authority further reduced under the new policy lines. In Liesnard's eyes, the situation was simply due to the lack of European supervision, caused in turn by personnel reductions, and, paradoxically, to Vermeulen's policy of creating "model villages," which had led to the rebuilding of the Musuumb a few years earlier. As the district commissioner saw it, "the population at Musumba has imagined that, once the capital was built, all it would have to do was to live off the 'basendji' in the bush who could be exploited at will!"[48] Stereotypes were called to the rescue:

> The Lunda, who are more intelligent than the Tshiokwe and also more receptive, are more easily won over to our ideas, but their conversion is not as frank, as witnessed by the passive resistance, filled with outward respect, which the occupying power meets on the part of their central authorities. This attitude of the Mwata Yamvo and his court is not exclusive to Sandoa territory. Wherever we have to deal with an important native authority, we experience difficulties. . . . As an example, I will mention [Mwami] Musinga [of Rwanda], who had to be deposed by the government of our mandated territories.[49]

To the south, the administrateur de territoire for Malonga, like so many field officials before him, argued the case for "his" chiefs and insisted that the abolition of subchiefs demanded by the Decree of 5 December 1933 should not lead to a restructuring of the Lunda state.

> Out of the 19 units in the territoire, 15 are organized as sous-chefferies depending from the Mwata Yamvo. It is impossible to regard Chief Mwata Yamvo as the sole intermediary between these communities and the European authorities, as section 27 of the new decree would have it. The huge expanse of Lunda country, and the very principles of its traditional organization prevent its unification.

And elsewhere:

> The Lunda sous-chefferies in Malonga territory are perfectly balanced administrative units. Supressing them radically is impossible, not only because this would completely disorganize our whole administrative system, but also because the huge empire of the Mwata Yamvo was never unified, even at the time of its greatest splendor.
> Regional autonomy . . . has always been the rule. [Chiefs like] Kazembe or Musokantanda always administered their people independently. . . . Only the more or less regular sending of a tribute linked them to the Musumba and to the Mwata Yamvo. This tribute was a mark of respect toward the man who was regarded as keeper of the ancestral land and did not symbolize political subordination or vassalage. In the event of the death of one of those great feuda-

tories [the Mwata Yamvo] did not even have the right to intervene and to cause the successor of his choice to be appointed: Musokantanda, Kazembe, and others have their "Tubungu" or electors, just as the Mwata-Yamvo himself.[50]

The debatable validity of these arguments (suggested, inter alia, by the repeated reference to the somewhat exceptional cases of Kazembe and Musokantanda) is less important than the fact that they were wholeheartedly endorsed by the district commissioner, whose own conclusion was that "the groups currently organized as sous-chefferies will have to be considered as chefferies under the terms of the new decree, which does not exclude that they might continue to maintain certain customary links with Musumba and the Mwata Yamvo."[51] And the provincial authorities in turn conceded: "This will indeed be the only rational solution if the principle that sous-chefferies must be abolished is not mitigated."[52] Nevertheless, they referred the matter for further consideration to Léopoldville, where it was brought up for discussion during the 1935 session of the Conseil de Gouvernement, a consultative body made up at that time of the provincial commissioners (i.e., governors) and other senior officials. Students of the bureaucratic phenomenon will appreciate the elegant casuistry which the governor general displayed on that occasion in order "to maintain the Mwata Yamvo's traditional authority while recognizing as chefferies the subdivisions administered by his subordinates." Quoting from the commission report which had formed the basis of the decree of 5 December 1933, the governor argued:

> When, according to custom, certain attributions are exercised within a given community by authorities [other than the chief] to whom custom allocates certain prerogatives, even if exclusively, these attributions can, and indeed should continue to be exercised by the traditional authorities so designated.

Of course, the governor general admitted, the authors of the report had probably intended their remarks to apply to traditional authorities operating within the area of the chief's jurisdiction, "but there seems to be no reason why the reverse should not equally be true. The mere fact that the European administration recognizes a given native authority . . . does not necessarily release it from its customary obligations toward another, higher traditional authority such as the Mwata Yamvo." The Lunda paramount, he concluded, "is still entitled to traditional tithes, which can be replaced by a charge supported by the native treasuries."[53]

After this piece of bureaucratic juggling, the way was paved for the official dismemberment of the Lunda state. During the course of 1936 all thirteen subchiefs of Malonga territory, as well as the seven sub-chiefs of Sandoa territory, were elevated to the rank of chiefs.[54] The two enclaved subchiefs of Kapanga and Tshibingu were absorbed into the Mwaant Yaav's own chefferie.[55]

As might have been anticipated, however, difficulties soon developed over the issue of the payments to be made by the newly elevated chiefs in lieu of tribute to their traditional suzerain, and a scenario reminiscent of the fate of the Cokwe tribute during the late 1920s began to unfold almost at once.

Upon learning of the compromise outlined by the governor general, the district commissioner objected that it would be unfair to have the impecunious native treasuries support the payments to be made to the Mwaant Yaav, since the newly elevated chiefs would receive salary increases, and suggested instead that these chiefs rebate 30 percent of the quarterly stipend they received from the administration (as opposed to their traditional benefits and to the rebates from the proceeds of the native tax allocated to them as a premium for effective tax collection).[56] The proposed formula was accepted by the provincial authorities, but its actual implementation lagged for another year, at the end of which the provincial commissioner (i.e., governor) reminded his subordinate that the 30 percent rebate was intended to be "in the nature of a tribute toward the Mwata Yamvo" and that the former subchiefs' traditional "condition of vassalage should be brought home to them each time they made a remittance."[57]

As a result of this intervention the rebates were effectively paid through the late 1930s, but from the very beginning the administrateurs for Sandoa and Malonga expressed strong reservations concerning the whole principle of tribute. The administrator for Sandoa wrote:

Tribute implies primarily a recognition of vassalage, but also, in principle, a recognition of services rendered by the chief to the community, in the form of administrative tutelage, military protection, and the maintenance of social order. While the first condition continues to be met in various degrees, it is legitimate to wonder if tribute still represents a counterpart for effective services. The

Mwata Yamvo's intervention in the administration of his former sub-chiefs is absolutely non-existent. He cannot be credited with a single measure, decision or advice, spontaneous or otherwise, to assist in the advancement of our program in his former sous-chef-feries. . . . His role in settling chiefly successions is mostly inopportune and seldom positive.[58]

His colleague at Malonga took a similar position and, observing that he had always argued against the "iniquity" of the tribute, he too stated that

the Mwata Yamvo's authority over the areas [under my jurisdiction] is non-existent. His past interventions in the administrative processes in this region were possible only because of our support, or under pressure from us; the authority he used to enjoy over the mass of the population was merely a function of the backing extended to him, in a rather high-handed fashion, by the Europeans.[59]

While expressing reservations concerning the situation in Sandoa territory, the district commissioner fully endorsed the views of the administrator for Malonga and suggested that a number of the former subchiefs be relieved of the obligation to rebate a portion of their stipends to the Mwaant Yaav since, as he put it, "his authority is really ignored . . . and his relations with the population are non-existent."[60]

This gradual disaffection with the Lunda paramount on the part of Belgian field officials — in marked contrast to the relative partiality exhibited by so many of their predecessors — reflected in part the colonial administration's growing preoccupation with economic problems, rather than with establishing the conditions of a stable colonial "order," but also their impatience with the Lunda court's increasing inability (or unwillingness) to perform as expected in this changing context. This latter phenomenon was not limited to the Lunda royal court, of course; between 1934 and 1939 no fewer than thirteen of the twenty-nine chiefs of Sandoa and Malonga territories were deposed, deported, or deprived of recognition for reasons ranging from an involvement in the spread of *Ukanga* to "incompetence" and "senility," and the turnover in other parts of Katanga was of the same order of magnitude. The sense of insecurity and frustration which, during the early years of the depression, gripped the European population (whose numbers dropped from 25,700 in 1930 to 17,600 in 1934)[61] also affected the chiefly class and was reflected not only in the dramatic spread of adventist and *bwanga*-centered movements but also in the appearance of serious feuds within the Lunda royal court.

Much of the tension which now mounted at the Musuumb had been initially generated by the news of the impending reorganization of the native administration structure. The prospect of seeing the Lunda subchiefs elevated to the rank of chief was viewed in Kaumba's entourage as a disturbing and mortifying development, not only because it represented one more step in the disruption of the ailing Lunda state, but also because the possible disruption of the tributary system threatened the whole intricate structure of the royal capital as well as its livelihood. Nor could Belgian promises to safeguard the tributary relationship between the court and the former subchiefs offer much reassurance after the precedent of the Cokwe tribute which, as the Lunda titleholders remembered only too well, had gradually been eliminated once the Cokwe areas had been organized into autonomous chefferies.

Kaumba's apparent subservience to the Belgian authorities in both instances fueled the bitterness of certain members of the royal entourage. In May 1935 a coterie of Lunda titleholders and court officials led by the Kanampumba (the titleholder acting as interim or "mock" king during an interregnum) attempted to discredit the Mwaant Yaav in the eyes of the colonial administration by exposing a number of irregularities committed by Kaumba (ivory poaching, hunting-code violations, etc.) and demanded that he be deposed for assorted instances of alleged misrule. Despite the rather low esteem in which Kaumba seems to have been held at that point by a number of Belgian officials, the district commissioner apparently felt that the status quo was preferable to opening the Pandora's box of a succession feud, and he forced the opposition clique to yield obeisance to the Mwaant Yaav.[62] All the same, the Mwaant Yaav's prestige was seriously damaged by this episode, and tension continued to smolder at the Musuumb. The Kanampumba eventually had to be deported in 1937 for continuing to foment intrigues against Kaumba. But while the colonial authorities might support the Mwaant Yaav against disgruntled members of his court, there was no doubt that they intended to treat the Lunda paramount no differently from other chiefs. Kaumba's *Nswaan Mulopw* (i.e., perpetual son and nominal successor-designate) was severely upbraided for his involvement in the diffusion of *Ukanga* in 1936 and when, in 1939, the Mwaant Yaav himself was implicated, as we have seen, the provincial commissioner personally deprived him of the right to wear his paramount chief's medal and warned him in no uncertain terms that he would not be given a third chance.

Thus, on the eve of Belgium's entry into the war, the colonial system had finally reduced the Lunda state to a size compatible with its own internal logic. Perhaps without having specifically intended at any given point to dismember the "empire," the Belgian authorities had deliberately ignored or treated as counterproductive any aspect of the traditional political process which did not visibly or immediately serve their policy goals of political quiescence combined with economic mobilization. The economic crisis of the 1930s, with its repercussions on the size and effectiveness of the European administrative super-structure, reinforced the tendency, inherent in all imperial powers, to regard native chiefs at least in part as auxiliaries of the colonial field administration.

Beginning in 1930, chiefs were increasingly commissioned as tax collectors, and by the time of World War II they were garnering nearly two-thirds of the proceeds of the native tax (see Table 10). The Decree of December 5, 1933, also generalized the system of compulsory farming of certain crops[63] and made the chiefs responsible for the per-

Table 10. District of Lualaba: native tax proceeds (in francs), 1935-1945.

Year	Total proceeds	Amount collected by chiefs	%
1935	2,800,000	1,337,057	47.8
1936	3,538,758	1,772,306	50.1
1937	4,768,689	2,269,170	47.6
1938	5,940,895	3,201,150	54
1939	6,306,572	3,589,581	56
1940	6,857,729	4,300,987	62
1941	7,291,561	4,715,566	65
1942	7,890,351	4,586,983	63
1943	8,391,837	5,184,704	61.7
1944	9,388,696	6,179,086	65.8
1945	9,707,254	6,311,455	65

Source: District du Lualaba, Rapports AIMO 1935-1945.

formance of these obligations which, although they were officially destined to promote the "economic education" of the peasants, were neither remunerated nor underpinned by governmental price support (in consideration of which farmers were "permitted" *not* to sell their crops if they failed to fetch a suitable price). Under the terms of the decree, the compulsory cultivation of cotton was introduced in Lunda in 1935, although production did not reach significant levels until the war.

Indeed, if the depression had intensified what came to be referred to as the *fonctionnarisation* of Congolese chiefs, it was the war effort that completed the process.

6

Out of the Depression and through
the War:
The Bureaucratizing of the Chieftaincy

Even though Lunda had been only marginally affected by the rapid economic growth of the Katanga copperbelt, it suffered many of the effects of the depression, not only in the form of a reverse flow of migrant laborers, but also because its meager agricultural production (accounting for less than one percent of the Congo's agricultural exports) was unable to countenance the tremendous decline of agricultural prices, which a Belgian official described graphically by noting that the Congo had earned more from the sale of 3,400 tons of native rubber in 1910 than from the export of nearly 200,000 tons of nonmineral produce in 1934.[1] But while the mining industry began to recover from the effects of the depression after 1933, it was not until World War II that Lunda was able to regain even the mediocre level of economic activity it had achieved in 1930. The only encouraging sign during the lean years of the mid-thirties was the development of a modest amount of mining activity and prospecting by the Forminière corporation. This provided three-quarters of the salaried jobs available in Sandoa territory and over half of those in Malonga during 1935, but, by 1936, that firm had completely abandoned its operations in Sandoa territory and substantially reduced the number of its employees in Malonga, where the railroad alone continued to provide a small number of jobs. In 1937 the number of African wage-earners in the territoire of Sandoa, which now covered the central and northern portions of Lunda, fell to the incredibly low level of 24 out of a total population of over 110,000 (Table 11). Under such conditions, it is hardly surprising to find that in 1936 the average yearly income of an

Table 11. District of Lualaba: number of African wage-earners,
by territoire, 1935-1945.

Year	Sandoa	Malonga	Kamina	Bukama	Kabongo	J'ville	Kolwezi[a]
1935	1,095	2,800	2,125	4,450	112	8,580	
1936	929	1,793	3,649	6,264	380	10,010	
1937	24	2,214	2,216	3,484	462	5,994	
1938	814	2,956	3,351	6,137	506	16,602	
1939	2,240	3,214	3,971	5,398	410	16,511	
1940	2,744	2,342	4,744	4,174	1,099	14,529	4,494
1941	876	2,628	6,154	4,427	1,113	15,453	4,494
1942	1,886	2,052	6,479	7,669	1,407	18,900	8,569
1943	1,840	1,603	7,202	6,196	1,437	22,097	8,876
1944	3,014	1,824	7,459	6,237	1,682	22,440	8,253
1945	3,206	2,307	7,454	4,581	1,187	23,908	7,972

Source: District du Lualaba, Rapports AIMO 1935-1945.
[a]Created in 1940.

African taxpayer in Sandoa territory was still less than 100 francs (the lowest in the entire district), while African employees of the Union Minière were earning as much as 560 francs at Jadotville and Kolwezi (see Table 12). On such a paltry income the native tax for residents of Sandoa territory was assessed at a flat rate of 32 francs, which amount, although lower than in other portions of the district, nevertheless represented one-third of the average income for the area. All the same, the unperturbed district commissioner estimated that "the tax rate remains within the limits of the natives' fiscal capability" and observed that in future years economic recovery would "inevitably" increase the taxpayers' monetary resources.[2]

The district commissioner did not specify, however, where these monetary resources might be found in a region where salaried employment was available to less than 5 percent of the adult male population and where export agriculture had failed so disastrously. As for the economic perspectives opened up by the introduction of cotton in Lunda, they were still largely conjectural. During 1934 able-bodied males in selected areas of Sandoa and Malonga territories were required for the first time to plant one-half acre of cotton, but

Table 12. District of Lualaba: average income brackets (in francs), per taxpayer, 1935-1944.

Year	Sandoa	Malonga	Kamina	Kabongo	Bukama	Jadotville	Kolwezi[a]
1935	60-85.30	82	107-177	71.4	220	108-158.80	
1936	91.87-99.58	163.10	145.74-167.12	140.50	184.62-207.10	190-560	
1937	106.16-127.16	168.29	200.08	170.33	227.50-266.00	230.45-480	
1938	136-154	195.4-202.4	204.75	180.25	225.33-540.00	233.33-600	
1939	134.33-146.66	190.16-196	200.6-224	193.66	240.91-540.00	256.66-600	
1940	178-182	235-244	240-267	234.5-240	268.33-540.00	245.00-600	
1941	190	235-244	240-278	240-246	269.50-540.00	245.00-600	226-600
1942	300	297-348	355-360	290-360	347-800	285-900	286-900
1943	322-382	340-360	400-420	324-400	380-850	315-2400	384-2400
1944	481-573.30	457.50-549.75	480-588	409-540	415-1200	339.5-2400	343-2400

Source: District du Lualaba, Rapports AIMO 1935-1944.
[a]Created in 1940

compulsory cultivation was still resorted to predominantly to ensure a steady production of food, rather than of cash crops.[3] The system was gradually extended, and, by the end of the decade, out of a "tribal" population of 117,323 adult males for the entire district, 90,121 were subjected to some form of compulsory cultivation, which for 75,981 of them included the planting of cotton.

With Belgium's entry into the war in May 1940, the Congolese population was subjected to unprecedented and often contradictory pressures. On the one hand, rural communities were called upon to fill the manpower needs generated by the war effort in the industrial sector, while being expected on the other hand to intensify their agricultural production for local consumption as well as for manufacturing and export. Yet, for every man sent to the mines or factories in Jadotville, Kolwezi, and Elisabethville — or even as far away as Maniema — the productive capacity of the villages, already sapped by the intensive recruitment of the 1920s, was bound to be further undermined. From 1940 through the end of 1944 the number of able-bodied males recruited from the District of Lualaba totaled 12,456, 15 percent of whom were sent outside Katanga. Even when repatriations are taken into account, the net loss of adult men by the rural communities of the district over these five years still amounts to 10, 651 — over 9 percent of the total number of adult male villagers in 1939, and an even more serious drain on rural manpower if we remember that only young men in good physical condition were being hired for industrial labor.

In the Lunda heartland these trends were even more pronounced. The number of adult males living in traditional communities in the two territoires of Sandoa and Malonga declined from 48,124 in 1939 to 41,218 in 1944. This represents a drop of 14.3 percent which, if factored to take into account the selectivity of labor recruitment, indicates that one-fourth to one-third of the young men of Lunda left their traditional homes during the war years.[4] A comparison of Tables 11 and 13 shows us where they went: to Elisabethville, of course, but mostly to the mining regions of the western Copperbelt around Jadotville and Kolwezi, where the migrant or "detribalized" population increased from 34,267 in 1939 to 72,862 in 1944 (an increase of 113 percent). Within Lunda itself the number of salaried jobs fluctuated between 3,500 and 5,500. These were largely filled by local people, especially in the central and northern regions where less than 0.5 percent of the population was living outside traditional communities in

Table 13. District of Lualaba: tribal and nontribal population, by territoire, 1934-1944.

Year	Sandoa	Malonga	Kamina	Bukama	Kabongo	J'ville	Kolwezi	Total
1934								
T	109,996	78,536	66,861	46,071	79,420	46,585		427,469
NT	834	3,429	3,020	4,600	----	19,734		31,617
	110,830	81,965	69,881	50,671	79,420	66,319		459,086
1939								
T	100,042	76,615	66,784	54,925	72,476	47,030		427,872
NT	185	3,333	7,507	5,177	786	34,267		51,255
	110,227	79,948	74,291	60,102	73,262	81,297		479,127
1944								
T	100,674	67,475	62,157	49,226	68,804	34,967	21,797	405,100
NT	436	5,011	9,684	11,127	238	55,752	17,110	99,458
	101,110	72,486	71,841	60,353	69,042	90,719	38,907	504,558

Source: District du Lualaba, Rapports AIMO 1934, 1944.

1944, although some 13 percent of the men were engaged in some sort of salaried occupation during that year.

The rapid decline of the adult male population of rural communities during the war years placed an exorbitant burden on all those who remained in the villages. The number of men subjected to compulsory cultivation was reduced from 90,121 in 1939 to 76,058 in 1944. The planting of cotton had been imposed on 75,981 farmers on the eve of the war; by 1944 that number had declined to 60,551 — a drop of some 20 percent.[5] To compensate for this manpower lag, the Belgian administration simply doubled the number of days a farmer was required to devote annually to his mandatory crops, bringing it to the rather staggering total of 120 days. The acreage to be kept under cotton was increased to one and one-half acres, but a few overzealous officials did not hesitate to "persuade" villagers to extend it to two and one-half acres.[6] And so it was that, despite the striking decline in the number of farmers, the District of Lualaba managed to increase its agricultural production from 23,204 metric tons in 1939 to a peak of 40,208 tons in 1943, a rise of over 73 percent. On a per capita basis, these figures actually represent a doubling of the agricultural output in four years, from 257.4 kilograms of produce per planter in 1939 to 515.5 kilograms in 1943.[7] The production of cotton in the district rose from 6,307 tons in 1939 to 9,965 tons in 1943 and that of manioc went from 10,961 tons to 18,133 tons over the same period of time; for the individual planter these represented increases of 83 and 91 percent respectively over their 1939 output.

Perhaps the most characteristic data, and certainly the most spectacular, have to do with the wartime resumption of rubber production. The collection of native rubber had been abandoned for a whole generation in most areas of the Congo and many villages had even forgotten how to prepare it for marketing. The initial phase of the war in Europe briefly stimulated demand, and production from Lualaba grew from a mere 2 tons in 1939 to 13.6 tons in 1940, only to be abandoned again in the following year. When Southeast Asia and Indonesia fell into Japanese hands, however, a fresh demand appeared and production from the district reached the unprecedented levels of 418.4 and 429.5 tons in 1943 and 1944 respectively. Except for the institutionalization of coercion, the hardships inherent in the collection of rubber had not changed greatly since the days of the Free State. One observer described the plight of Katanga villagers enjoined to "make rubber":

Have you seen those villagers from fertile areas, their fields covered with cotton in bloom, leaving for the forest dragging with them their wives, their children and their small cattle, covering up to sixty kilometers, then . . . returning after five or six weeks, often diseased, having consumed their chickens, their goats and their sheep, or having bartered them for maize or manioc with the natives who live on the edge of the forest but do not have to collect rubber simply because they are from another territoire?[8]

An equally oppressive but even more generalized burden on the resources of rural communities was the vastly increased amount of roadwork. During the first three years of the war, nearly 1,000 miles of new roads were built in the District of Lualaba. The total road network in the district grew from 5,300 miles in 1939 to 6,535 miles in 1944, but only 888 miles of roads were maintained by the government: the rest was the responsibility of the various native communities. In the cotton-growing areas, such as Lunda, roadwork was financed by the proceeds from cotton sales as determined by the government marketing board, which means that the labor performed on the roads by the villagers was being "remunerated" with the money they should have been earning from their cotton. Technically, the development of new roads was supposed to reduce the need for human portage, but the limited number of trading points and the lack of transportation equipment prevented the system from operating as planned, especially where food crops were concerned, and it was not uncommon for villagers to spend a total of two months each year carrying goods to market without remuneration, even though their villages had been made accessible to trucks through their own labor.[9] Road maintenance itself resulted in a different form of portage. The chief sanitation officer for the Union Minière observed:

Throughout my entire colonial career, which already numbers many years, I have never seen as many natives on portage duty as I do nowadays: everyone, including women and children, is carrying loads of earth along the roads to replace what has been lost and will soon be lost again, in the form of mud during the rainy season and in the form of dust during the dry season.[10]

But the real price of the war was also being paid in more indirect ways. The invidious contrast between the relative ease and security of life in the city and the deadening drudgery of village life demoralized the rural population more than any previous encroachment on traditional societies. Leopold's agents might have been more deliberately brutal, but then the forest was the only place to run to. Now there was the city, the well-ordered camps run by the Union Minière and proudly shown as models throughout southern Africa. In the "bush," by contrast, medical services had virtually disappeared, schools were more inadequate than ever, and goods were scarce.[11]

Whatever they had heard about the urban centers (and what they heard, more often than not, was positive), the peasants knew what their own plight was like: an endless round of unremunerated *corvées* and unrewarding labors with no visible end or purpose under the vexatious authority of a harried band of native auxiliaries. Production schedules imposed by the administration left no time for fishing or hunting (the latter being also hampered by the lack of gunpowder), thus eliminating a major source of proteins in the rural diet. Even more important, no time was left for the elaborate array of social obligations centered on births, marriage, and deaths, or for community functions such as palavers, litigation, dances, etc. Houses, wells, and fences went unrepaired. The quality of village life declined sharply during these years, thus causing further demoralization and reinforcing, by contrast, the attractiveness of migration. It is significant that the rural exodus of the 1940s took place in the absence of any large-scale recruitment drives such as those that had occurred in the twenties: villagers were now escaping, rather than being herded off, to the distant cities.

Women increasingly took the place of men in the fields and even children were pressed into service (for rubber collection notably), to the detriment of their already rudimentary schooling.[12] Traditional family life was seriously disrupted by these various developments, and the number of marriages in rural areas declined substantially, as did the rate of natural population increase which, in the northern and central areas of Lunda, had already sagged during the 1934-1939 period to the rather unencouraging yearly average of 2.59 per thousand (versus 5.03 in the south).

Despite, or perhaps because of, this devastating erosion of traditional society, there was little explicit protest in the rural areas during this brutalizing decade, other than an increasing resort to compensatory escape mechanisms, such as those represented by adventist beliefs and propitiatory cults. As mentioned earlier, *Ukanga* remained popular in Lunda throughout the war and, apart from sporadic reactions by the administration, such as the dismissal of Chief Lumanga in 1943, Belgian officials had little time to check its diffusion.[13] The resurgence of an ancient superstitious fear, that of the "Mitumbula," may also have been an indirect expression of the latent anxiety caused in village societies by the continuing depletion of their human potential: the Mitumbula were accused of kidnapping villagers for various sinister purposes, notably to supply certain Europeans with human flesh.[14]

Kitawala, on the other hand, failed to make any significant inroads into Lunda and the fairly widespread adventist beliefs which appeared in connection with specific aspects of the second world war (such as the anticipated arrival of Black American troops[15] or Simon Mpadi's prophecies regarding Hitler's role as a potential liberator) had no visible impact in this remote and isolated region, although they had found an audience in the Kolwezi area.[16] As a matter of fact, the only adventist movement worthy of note, that of Ilunga Clément (or "Kelema"), made its appearance in the upper Kasai valley five years after the war and was in many ways a repetition of the ancestor-return theme of 1931.[17]

In neighboring areas, meanwhile, dissatisfaction was being expressed in more urgent ways. Uncoordinated strikes and unrest were practically endemic on the Katanga copperbelt from 1941 to 1943, and in one murderous confrontation at Elisabethville in December 1941 some seventy miners were shot by security forces. At the center of these tensions were salary demands and the growing frustration felt by skilled Africans at the enormous wage gap between themselves and their nearest European counterparts. The comments reportedly made by several miners after the 1941 massacre, "Banafwila mali yabo" (They died for their wages),[18] illustrates the growing awareness and solidarity of the industrial workers, as well as the specific nature of their grievances which were met, at least in part, by substantial salary raises: from 1941 to 1943 the range of industrial salaries in the mining areas was raised from 30-50 francs to 50-200 francs a month.

Of the several violent incidents that broke out during the last year of the war in various parts of the Congo, only the so-called Luluabourg mutiny of February 1944 had any serious impact in Katanga. Actually, the mutiny itself, which had been planned for over a year, was scheduled to coincide with demonstrations by white-collar workers at Elisabethville, but these were forestalled by a wave of arrests. When the troops at Luluabourg did mutiny some time later, they were widely followed by the local population, and there were also brief uprisings by the detachments stationed at Jadotville and Kamina.[19] After the movement had been suppressed, the Luluabourg mutineers, like their Batetela predecessors half a century earlier, dispersed into a number of small bands, several of which wended their way into Katanga, sparking a trail of minor incidents in the Luba regions of the province. One such party made its way across the lands of the Mwaant Yaav and, after defeating a small detachment of loyal troops at Samabanda (west of Kapanga), they crossed the Kasai into Angola where they were promptly captured by the Portuguese, who turned them over to the Belgian authorities.[20] Lunda villagers apparently sympathized with the mutineers and offered them food and guides (possibly under duress), but showed no signs of wanting to join in an insurrection.[21] The whole episode left no visible traces, and calm, born from resignation rather than acceptance, soon returned to the rural areas.

The end of the war brought no immediate relief from the constraints of the wartime years. Production norms were eased or abolished, but prices remained high and manufactured articles continued to be scarce. The district commissioner for Lualaba, writing in early 1946, commented:

> Now that peace is back, [the natives] wonder at the slowness of the return to normalcy, to abundance and low prices. In order to stimulate the war effort, we had quite rightly blamed enemy activity for supply problems, for the high price of clothing and other common articles. The cause has disappeared, but the effects linger on and this is what the natives find rather odd.[22]

The villagers had cause to be disappointed. The wartime development of cash crops had expanded the base of the monetary economy. The volume of currency in circulation more than doubled during the war. The annual income of an average taxpayer in Lunda, which had varied from 134 to 196 francs in 1939 now ranged from a low of 457.50 francs to a possible high of 573.50 francs. Wartime shortages had curtailed purchases and resulted in a limited amount of monetary accumulation in African hands. Much to their dismay, however, the farmers now discovered that their money would not buy even one-fifth of what it would have bought ten years earlier. From a base of 100 in 1935, the African cost of living index had risen to 545 in 1945. If we apply this indicator to the scale of average incomes presented in Table 12, we find that the median annual income per taxpayer in Sandoa and Malonga territories had progressed, in real terms, from 77.3 francs in 1935 to 94.6 francs at the end of the war — hardly a spectacular improvement when set against the human cost of that brutal decade.

Traditional authorities suffered as much as the whole rural society from the constraints generated by the depression and the war. Theirs was a particularly unenviable task, however. Caught between their people and the demands of overworked field officials, they found themselves faced with the discouraging alternative of performing the role of mouthpieces for the European administration or being bypassed by the lowliest native auxiliaries of the colonial bureaucracy. Antoine Rubbens, an Elisabethville lawyer active in progressive Catholic circles, gives a vivid impression of the chief's awkward position when he describes him "circling like a sheepdog" around the native *moniteur* who is rounding up his subjects for the morning roll-call before herding them off to the fields to work on their government-imposed crops.[23] By creating a single abstract class of native administration units, the *circonscriptions indigènes,* the Decree of 5 December 1933 had deliberately chosen to minimize the distinction between traditional units (chefferies) and the artificially created secteurs. By the same token, the same legal status and standards of performance were increasingly applied to those chiefs who derived their authority from the internal dynamics of native society and to those who owed their position primarily to a government warrant of appointment. Chiefs had of course long been expected to relay administrative directives to their subjects, but it was often tacitly assumed by field officials that, in the absence of representative institutions, the chiefs were qualified to speak for their people, as it were, and to express the preoccupations of the community. This "hidden face" of the relationship between native chief and colonial official was now increasingly ignored in favor of a streamlined, hierarchical pattern in which chiefs (whether traditional or not) were regarded merely as the ultimate link in a single chain of command extending from Brussels down to the remotest hamlet. As a result, traditional chiefs gradually lost what little capability they had to shield their communities from the increasingly exacting demands of the administration. Rubbens who, as a Catholic layman, shared many of the reservations felt toward the idea of native administration by a

majority of missionary circles, summed up the chiefs' dilemma as follows:

> Too often, the chiefs have neither the courage to explain to the of-
> ficials the increasing burden which the accumulation of chores
> represents for their subjects, nor the strength to divide them equita-
> bly and to have them accepted. . . . They expose themselves to re-
> buffs from their best subjects by proclaiming ever more oppressive
> orders without understanding them and without having the strength
> to have them respected; while in the eyes of the whites (such is the
> spirit of the Decree) they appear to be accomplices of the natives'
> apathy.[24]

The administration itself could hardly be expected to assent to the
terms of this indictment but, while insisting that the chiefs' collabora-
tion was effective and fruitful, some officials recognized that

> this assistance involves risks and dangers for which they should be
> compensated. Natives tend to regard a chief who is loyal to the
> administration as our lackey, rather than as the outstanding repre-
> sentative of traditional authority. The consequence is a disaffection
> for which nothing can compensate.[25]

What many of them felt privately, Rubbens suggested, may have been
somewhat different:

> Field officials do not like their job any more. Indeed, it is not a nice
> job they are being forced to do: misusing the authority and confi-
> dence they have acquired to disperse families, deport laborers and
> ruin villages in the name of production, and no longer having the
> time—the shame of it!—to hear the palavers of "their" natives.[26]

In fact, this sense of demoralization went back to the 1933 reorganization. The density and quality of the field administration had suffered the effects of the economy drive inaugurated at that time and
had never fully recovered from a reform which had slashed the
effectives of *la territoriale* (as it was familiarly known) without reduc-
ing their responsibilities. The emphasis on budgetary economy and
increased productivity which characterized the depression and
wartime years had resulted in the proliferation of specialized officials
(treasury clerks, crops officers, etc.), whose numbers swelled the depleted ranks of the colonial service but who could not fill the role of the
general-purpose *administrateur* when it came to maintaining contact

with the rural population. Answering directly to their respective service branch at the provincial level, they were often resented by the
generalists — particularly the crops officers, who were empowered to
give direct orders to village chiefs, either directly or through native
moniteurs, thus undercutting the local administrator's authority over
"his" villagers.

The fragmentation of local European authority, combined with the
unrelenting pressure for economic performance, led to a number of
partly improvised attempts to achieve a greater responsiveness of the
native administration structure to governmental stimulations. The
native administration Decree of 1933 had of course been the most
comprehensive and systematic effort in this direction, but other related schemes were also resorted to, often on a tentative, experimental
basis. One such program involved the increasing resort to nontraditional councils of notables, partly to generate broader acceptance of
administration policies and partly to counteract the chiefs' inability —
or unwillingness — to assume nontraditional responsibilities. The success met by these councils varied considerably over the years and
according to local circumstances. Apart from their rather uncon-
vincing record as trouble-shooting bodies, however, these councils
achieved little and suffered from the lack of a clear purpose. One
potentially significant role which they might conceivably have played,
that of providing for an embryonic form of local representation, was
never effectively developed nor, it seems, systematically pursued. In
1938 an attempt to introduce younger members into the councils in
the Districts of Lualaba failed against the determined opposition of
the older notables and was never again repeated.[27] It was not until the
late 1950s (specifically, until the Decree of May 10, 1957) that a belated and rather ineffectual attempt was made to introduce a conciliar
system as an instrument of rural local government.

Another tentative scheme to turn traditional elites into more effective auxiliaries of European policy involved the systematic channeling
of the sons of chiefs and notables to the school system, limited as it was.
The idea was neither new nor original (other colonial powers had been
applying it for years), but at the end of 1934 the provincial commissioner (governor) issued specific instructions, outlining a three-stage
program. Sons of chiefs and notables would first receive basic literacy
training in a village school, then a full primary education in a mission
school, followed in turn by administrative training to be provided by

Belgian field officials.[28] The results of this program were not particularly impressive, however. In the Lunda heartland alone the number of sons of the traditional elite receiving some sort of schooling went up from 87 in 1935 to a peak of 377 in 1943, but the overwhelming majority of these (259 in 1943) were enrolled in two-year rural schools. The absence of any secondary schooling for Africans (apart from Catholic seminaries) sharply limited the potential political effects of this educational program. There was to be no equivalent in the Congo of the Ecole William-Ponty of French African fame until after the war, when the Jesuits opened a center for advanced studies at Kisantu (Lower Congo) which later became the nucleus of Lovanium University. As for the tutoring to be provided by Belgian administrators (*stage*), it was never systematized and appears to have been organized more or less on a volunteer basis. Interest in the program on the part of the overworked Belgian field officials was sporadic, to say the least, and clearly fluctuated in an inverse ratio to the attention they had to devote to their many other duties. Thus, out of 42 sons of chiefs and notables who went through the *stage* between 1935 and 1945 in Sandoa territory, 36 received their training in 1935, 1944, and 1945. In neighboring Malonga, a first trainee was accepted in 1938 and two more in 1939, after which local officials apparently abandoned the program altogether.[29] Nor was the lack of an appropriate training for future chiefs limited to Lunda alone: in 1949 in his annual report to the Belgian Parliament the minister for colonies admitted that "the low degree of sophistication of the chiefs, notables and councillors represents a source of major difficulties in the administration of many units and the training of these auxiliaries remains one of the most important tasks incumbent upon the field administrators."[30]

Whether or not this statement represented anything more than a pious wish, there is no doubt that such training programs made sense only to the extent that the administration was prepared to rely exclusively—or at least predominantly—on traditional authorities to implement its policy goals at the village level. In fact, as noted earlier, this determination was all but obvious during the depression and war years, as direct intervention by nontraditional agents (whether black or white) became increasingly frequent.

To the extent, however, that chiefs continued to be regarded as important relays in the administrative structure, there were other ways for the administration to ensure the gradual emergence of more "competent" chiefs, i.e., more responsive to the colonizer's values and priorities. One such method, which had been repeatedly if not systematically resorted to over the years, involved the intervention of Belgian officials in settling succession issues. Yet such crude tampering with traditional procedures was viewed by the Belgians as acceptable only in exceptional situations and, as patterns of native administration became increasingly stable, more subtle forms of intervention were resorted to. One such method was based on the idea of establishing a measure of predictability over the outcome of future successions. This involved a close scrutiny of chiefly genealogies, followed by consultations with traditional authorities with a view toward narrowing down the field of potential successors to a small number of mutually acceptable candidates who could then be watched (and trained) more effectively by the administration,[31] and perhaps even be delegated some duties before the incumbent chief's demise. Thus in 1945 the administration was able to secure the weeding out of all but two potential successors to the title of Muteba, and to limit to three names the field of candidates to the succession of Mbako. The same methods were applied the following year to Tshibamba, Lumanga, and Kayembe Mukulu. In the case of the Mwaant Yaav's position, however, although Kaumba was only too willing to intimate his own personal choice of a successor, genealogical claims were so tangled and potential rivalries so acute that the administration deemed it advisable not to push matters too far, leaving a field of some ten candidates which, the administrateur confidently asserted, "is bound to be even smaller when the actual succession opens, either because of the death of some candidates or for other reasons".[32]

Needless to say, the increasingly cavalier way in which chiefs were deposed for a variety of reasons after the 1933 reorganization of the native administration system made it possible to eliminate those who did not fit the Belgian authorities' norms of "competence." Colonial officials soon discovered, however, that dismissing a chief was not in itself a solution unless his successor, no matter how docile, possessed enough authority to secure the villagers' compliance with the objectives set by the administration. Thus although every one of the four Cokwe chiefs was replaced at least once during the 1930s, the quality of their successors' administration did not substantially improve—at least from the viewpoint of Belgian officials—and a report written in 1934 concluded somewhat despondently that "the lack of authority seems to be endemic among the Cokwe."[33] Among the Alwena, the native administration structure which, after years of fumbling, had finally been set up in 1932 threatened to collapse when, in 1936, one of the two chiefs of the Katende line again removed himself to Angola, as was the custom with members of this border tribe. The temptation for Belgian authorities to resolve such problems by merging more or less arbitrarily small or unmanageable native units into secteurs was mitigated only by the intricacy of the problems involved in selecting an acceptable chief for such an artificial entity. Thus, the recurrent idea (briefly applied in the 1920s) of merging the two small but densely populated chefferies of Tshisenge and Kandala into a single Cokwe sector was again revived at the end of the war and studied year after year until it was finally given up in 1954, after the Belgians had recognized that the chief with the greater amount of traditional prestige was also, from a European viewpoint, the more mediocre of the two.[34]

In the case of the Lunda chiefs, however, the high rate of attrition maintained by the Belgian authorities actually helped to reinforce the Mwaant Yaav's personal position for, even though they had no qualms about dismissing Lunda chiefs on a variety of nontraditional grounds, Belgian officials would invariably attempt to preserve a semblance of traditional continuity by consulting the Lunda paramount before appointing a successor. In this fashion, Kaumba was able to maintain a certain amount of leverage over members of the Lunda aristocracy by advancing the political fortunes of his relatives and supporters. Actually, the extent to which Kaumba's recommendations were followed by the colonial authorities is almost impossible to determine, but since the authorities themselves had an interest in preserving appearances, it was relatively easy for the Mwaant Yaav to persuade the Lunda notables in retrospect that the success or failure of their chiefly ambitions had been due to his intercession. In at least one instance, Kaumba's intervention in a succession case resulted in a visible gain for one of his protégés: when Chief Lumanga was deposed in 1943 because of his involvement in the *Ukanga* cult, the Mwaant Yaav saw to it that he was replaced by a member of his own entourage, Mutshaila Ditende.[35]

This type of rear-guard action waged by Kaumba and his entourage to consolidate their personal positions, even in the smallest way, was of course not new. In one form or another Kaumba had been faced with the more or less covert opposition of segments of the Lunda aristocracy ever since the day of his disputed election to the throne, and even though the colonial administration's backing had seen him through such crises as the 1935 cabal mounted against him by the Kanampumba, the same administration had also indirectly reduced the court's prestige — and income — by its reorganization of the native administration system. It is hardly surprising, therefore, that Kaumba and his entourage seized even the slightest opportunity to utilize whatever influence they could still wield in order to advance their interests and thus to arrest a further eroding of their power.

Such delaying tactics could take many forms. Playing off one faction against another, and more specifically, encouraging a local *mwaantaangaand* to pay direct allegiance to the court rather than to the chief recognized by the Belgian authorities (especially if this chief was himself a *mwaantaangaand*) had been a standard practice for Kaumba ever since the beginning of his reign and he continued using it to the end, even though the Belgians systematically refused to further fragment native administration units merely to satisfy the ambitions of local notables.[36] This did not deter Kaumba from granting unauthorized privileges to a *mwaantaangaand* (such as the right to maintain his own policemen and tax collectors) if this could help to undermine the authority of a Lunda chief who tended to forget his traditional obligations to the royal court.[37] The latter tendency had of course been enhanced by the 1935 elevation of all Lunda subchiefs to the rank of chief, particularly on the part of those who, like Kayembe Mukulu, Tshibamba or, before 1943, Lumanga, were descendants of local land chiefs rather than *yikeezy* appointed by the Mwaant Yaav.

The administration of customary law was another area where the court could maximize its influence — and, not unrelatedly, improve its material position. Belgian administrators recognized, somewhat ruefully, that a large number of cases were being decided in the Mwaant Yaav's court on the basis of interest or personal preference. Commenting on the outcome of a land dispute which Kaumba had adjudicated, with obvious partiality, in favor of his court physician, a Belgian official concluded:

> This proves once again that the Mwata Yamvo does not decide cases according to equity but rather seeks his own interest and that of his protégés. . . . As he always did and always will do, Chief Mwata Yamvo likes to intrigue to his own advantage, or to the advantage of his entourage and of his own favorites, in contravention of the most elementary forms of equity.[38]

And another official concurred: "Thanks to a well organized administrative hierarchy, [Kaumba] can afford to help his favorites and despoil the others."[39]

Whether he tried to cling to every remaining shred of royal
influence or merely attempted to retain his power of patronage (and
who could draw a clear line between the two?), Kaumba spent the last
years of his reign trying to resist any further decentralization of his
prerogatives. He insisted on having his own tribunal hear customary
litigation involving the subjects of Chief Tshibamba, rather than
authorize the latter to organize his own local court. He also objected
strenuously to the idea of having his own chefferie divided into territorial subgroups.

Ever since the time when Belgian authorities had first attempted to
organize native administration units in the Lunda heartland, the historical core of the former empire (extending from the lands of Mutombo Mukulu to the Kasai) had remained under the Mwaant Yaav's
direct jurisdiction, and the deliberate coincidence between the limits
of this royal domain and those of Kapanga territory had represented
an implicit recognition of the unique character of this vast chefferie.
After the administrative organization of 1934-1935, however, the situation had changed significantly: all Lunda subchiefs had been elevated to the rank of chief (thus placing them technically in the same
administrative category as the Mwaant Yaav), and the territory of
Kapanga itself had been deprived of its separate existence and attached to Sandoa. Thus, while it expected the former subchiefs to continue acknowledging the Mwaant Yaav as their customary suzerain,
the colonial administration, for its part, increasingly tended to treat
him simply as a chief having an unusually large number of subjects but
otherwise not very different from other native rulers.

In 1948, Kapanga territory recovered its separate identity after an
interim of fourteen years, but it immediately became apparent that
the administration's motive for stationing an administrateur de territoire near the Mwaant Yaav's capital was primarily to achieve a more
effective control over the Lunda court and to encourage the nascent
production of cash crops (cotton and peanuts). The internal organization of the royal domain was so intricate, however, that Belgian
officials soon found themselves despairing of its ability to reconvert
toward nontraditional pursuits. Traditionally, each major titleholder
of the court had been granted a sort of benefice in the form of a number of villages whose tributary payments provided him (or her) with a

permanent source of income. Instead of being located in a single region, however, villages paying tribute to a given titleholder were usually scattered into several noncontiguous areas, reflecting the historical stratification of privileges conceded by successive emperors, the fluctuating importance of certain titleholders over past generations, the perpetuation of ritual prerogatives predating the emergence of the Lunda state or the disruptions caused by the Cokwe invasions. Thus the Kanampumba and the Mwadi, for example, were each entitled to raise tribute from no fewer than four different groups of villages located at opposite ends of Kapanga territory, and the fiefs of other titleholders were similarly segmented into a number of noncontiguous clusters.

So long as the production of cash crops in the Lunda heartland had remained negligible, the colonial authorities had been content to rely on the Mwaant Yaav as their only intermediary in Kapanga territory, but with the growth of export agriculture during the late thirties and the forties, they increasingly complained that the Lunda paramount was neither able, nor indeed willing, to collaborate in the achievement of the administration's new policy goals. Nor could the titleholders, whose place and role was at the court, be effectively used as instruments of territorial decentralization. As a result, Belgian field officials found themselves having to deal, for an increasingly large number of matters, with a host of village headmen and land chiefs, none of whom controlled a sufficient number of people to warrant the creation of local administrative services. The first administrator stationed at Kapanga after the area had been restored to the rank of a territoire complained:

> The chefferie [of the Mwaant Yaav] which covers the whole territory is not subdivided into customary units of such size as to permit the organization of the necessary administrative agencies — responsible headmen, policemen, tax collectors, crops officers, agents of the *état civil*, etc. At this point, we are faced with a multitude of petty intermediaries: village chiefs, land chiefs, and our efforts are being dispersed. In fact, the only intermediary would be the Mwata Yamvo, who is not equal to the task. . . . The political organization of the chefferie conflicts, in actual practice, with any division into significant units. . . . [Titleholders] would make ideal intermediaries if their tributaries were collected into a single area, but such is not the case: they are scattered over the entire breadth of the chefferie.[40]

But the true dimensions of the problem, he added perceptively, were that

> the Mwata Yamvo would be totally disinclined to countenance the organization of administrative sub-units, even if these had a customary basis. He fears that such sizeable units might gradually seek to make themselves independent.[41]

At the same time, however, Belgian officials insisted that "the current centralized system is detrimental to the administration of this vast chefferie. Even if he was more cooperative, the Mwata Yamvo could not cope with the task at hand; he is, in any case, too old to be of any appreciable assistance".[42] "His only contribution," wrote another official, "resides in the fact that he does not obstruct our administration."[43] And yet, every report concurred, Kaumba's authority remained "undeniable," even though (as one of them noted), "he uses it mostly when his own interests are involved, and hardly ever when it comes to helping the administration."[44]

Successive Belgian officials noted with growing irritation the intense pride of the Lunda court and Kaumba's haughty reluctance to deal with such a lowly figure as an administrateur de territoire, let alone with the very junior agents who had been assigned to Kapanga during the years when the area was being administered from Sandoa, but they generally ascribed this attitude to the king's great age and to his court's anachronistic attachment to an antiquated tradition.[45] Time and the younger generation, they felt, were not on their side. Not that the officials had any great sympathy for the handful of local évolués (whom they disparagingly insisted on referring to as "évoluants"). Good broussards that they were, the local administrators had a deep-seated distrust for the "trousered African" and a patronizing fondness for the "real" Africa of chiefs and mud huts. The educated, on the other hand, were "more or less literate," "thirsting for independence," inclined to "snobbery" toward the older generation, and definitely in need of "constant surveillance."[46] To keep them busy, the administration had organized a social club at Kapanga, the "Cercle Lunda," numbering all of seventeen members drawn in nearly equal parts from the alumni of the Methodist and Catholic missions; but things did not go smoothly between the two denominations, and in 1948 the two boy-scout troops organized by the rival churches had actually come to blows, which had nearly wrecked the little club. Fortunately, the eldest

son of Joseph Kapenda Tshombe, the most prominent African businessman in the whole of Lunda, had used his genial influence on both
sides to bring about a reconciliation. Quite a capable fellow, that
Moïse: helping to run the Methodist boy-scout troop, organizing soccer
games . . . Of course, the court party was wary of the *évolués* and of
their social club which, they felt, was "directed against them"; but
"time," suggested an unwittingly prophetic official, "will make them
understand that their suspicions were not founded." Time would also
weaken the hold of the older generation which was "too superannuated
and too deeply anchored in their ancient routine for [the administra-
tion] to secure any improvement in their cooperativeness. This huge
chefferie will be set straight only when a new leadership takes over
from the old order."[47]

In the meantime, the administration was content to structure some
of its own activities in such a way as to be able to operate at a lower territorial level than that of the entire Mwaant Yaav chefferie. Thus, for
purposes of census taking and the collection of the government-imposed native tax, the chefferie was divided into a number of geo-
graphic areas, but this sort of territorial breakdown was of course
functionally specific and did not really affect the native administration
structure.[48] To the extent that such operations reflected European concerns, the Lunda court did not object to being bypassed and Kaumba
willingly rendered unto Caesar by delegating to local headmen the
unpopular task of collecting the native tax.

In 1950, however, the newly appointed acting administrateur for
Kapanga decided to go one decisive step further, and to organize these
precincts into veritable native administration units by setting up subordinate native courts in each one of them, and by consolidating a
number of administrative duties into the hands of local headmen.
Thirteen such subordinate jurisdictions had been established by the
end of 1950 and their combined activity at that date accounted for the
handling of 353 cases (against 199 in the Mwaant Yaav's own court at
the Musuumb). Although the total amount of court fees generated by
native courts in the territory increased from 40,265 francs in 1949 to
49,837 francs in 1950, the fees collected by the Mwaant Yaav's court
declined from 26,000 francs in 1949 to 15,455 francs the following
year.[49] This loss of revenue alone would have sufficed to cause an uproar at the royal capital, but the creation of the subordinate courts was
only one facet of a concerted effort to decentralize the Mwaant Yaav's

chefferie by increasing the responsibilities of certain local headmen.
The administrator for Kapanga admitted as much when he described
his policy moves:

> Actually, in order to circumvent a possible veto on the part of the
> Lunda paramount . . . the precinct courts were created first and
> placed under the most influential headmen; these were then subdel-
> egated to collect the native tax and finally resorted to for other
> duties, such as the relaying of instructions relative to [compulsory]
> crops, the recruiting of labor for construction projects, the super-
> vising and policing of marketplaces, etc.

The standard complaints raised against Kaumba and his entourage by
previous officials—prevarication, ineffectiveness, uncooperativeness,
etc.—were again brought up to justify the new policy: "It proves increasingly useless to work through his intermediary," he commented.
"Every method has been resorted to, but without any result."[50] Some of
the administrator's other comments, however, revealed more than a
passing irritation toward the old paramount and echoed some of the
strictures raised against the Lunda empire itself some thirty years
earlier:

> The Mwata Yamvo's attitude toward the Europeans is clearly hos-
> tile. . . . We can only gain by dealing firmly with him and by watch-
> ing him constantly. . . . Some field officials still feel that they should
> refer to him for the succession of chiefs from neighboring areas.
> This is, in my opinion, an error, the only consequence of which is to
> restore openly his autocratic rule over regions that have been de-
> tached from his chefferie many years ago. This manner of proceed-
> ing takes on, in the Mwata Yamvo's eyes, the appearance of having
> the administration defer to his will and weakens correspondingly the
> influence of local officials. Is there a greater paradox than to at-
> tempt forcing the authority of the Lunda paramount over popula-
> tions that defeated him several decades ago? . . . The myth of a vast
> Lunda empire cannot be revived without seriously harming the
> administration of that tribe—a task that will be made easier by the
> fragmentation of the Mwata Yamvo's ancient chefferie.[51]

Kaumba's reaction to this thinly veiled attempt to dismember the
last core of the Lunda empire was to repudiate, as overtly as he dared,
all forms of cooperation with Belgian officials. He refused to appear
during the district commissioner's tour of inspection, arguing that his

salary was too paltry for him to maintain the appropriate decorum for such an occasion. He repeatedly refused to instruct his subjects residing at the Musuumb to present themselves to Belgian health inspectors, and, when reprimanded about this, he replied with bitter irony: "In the old days I had the right to put to death anyone who disobeyed my orders. You took away this power and so this is now your responsibility."

The war of nerves between the Lunda court and the Belgian administrator went on through the latter part of 1950 and the early months of 1951. Watching the situation with more dispassionate eyes, however, the district and provincial authorities soon became concerned about the deleterious effects of a protracted conflict between the Lunda aristocracy and the administration. At the same time, the overzealous attitude of local headmen promptly made them thoroughly unpopular with the villagers. In March 1951 the impulsive administrator for Kapanga was transferred and his predecessor was brought back to soothe ruffled tempers and restore normal relations with the Musuumb. Local headmen were "reminded" that their authority was derived "from the Mwaant Yaav alone, that they were merely his representatives and collaborators in their respective areas, rather than the vexatious executive agents of the local Belgian authorities."[52]

Kaumba did not live to enjoy his vindication, however; he died on the twenty-eighth day of May, 1951, and was promptly succeeded by the 53-year-old Chief Mbako Ditende. In fact, Kaumba's death was timely, both from the viewpoint of the administration and from that of the Lunda aristocracy. Although they had supported him for the sake of continuity during the last crisis of his reign, the colonial authorities had long been awaiting the accession of a more "modern-minded" ruler who would be more responsive to their concepts of progress. The old Mwaant Yaav's demise also made it possible for the administration to gloss over the somewhat embarrassing outcome of the recent confrontation and conveniently to turn over a new leaf without loss of face to either side. Relations between the administration and the Lunda court would now take a fresh start, with a new and presumably more receptive sovereign being enthroned at the Musuumb. And, from the viewpoint of Lunda court circles, few men could be better qualified than Mbako Ditende to maintain a dialogue with the Belgian authorities. Like many members of his generation, Ditende had been

born, around 1898, of a Cokwe father, Tshipunda, and a captive
Lunda noblewoman, Matshika, who was a granddaughter of Mwaant
Yaav Naweej a Ditende and had held the title of *Nswaan Muruund*
(*Swana Mulunda*). Shortly after his birth, the Cokwe had been driven
out of the Lunda heartland and the Lunda captives had been freed,
along with their progeny. The young Ditende came to live at the court
of Muteba, whose accession to the throne in the place of Mushidi I had
been greatly helped, it will be recalled, by the officials of the Congo
Free State. Ditende, along with Joseph Kapenda, was one of the first
pupils of the newly established American Methodist mission but,
unlike Tshombe's father, he did not convert to Christianity until April
1960, three years before his death. His familiarity with the ways of the
whites made him a valuable intercessor between Muteba and the
colonial administration and he became the Mwaant Yaav's chief
spokesman and interpreter. In that capacity he accompanied the
Liégeois expedition which led to the founding of the post at Kafakumba. When the Belgian authorities attempted to restructure the
Lunda heartland under Muteba's authority, by inviting him to
appoint resident lieutenants for various areas, Ditende was rewarded
for his services by being made Chief Mbako (1917), in spite of his
youth. This was a key position since it enabled Ditende to keep in
constant contact with the district commissioner at the newly established headquarters of Sandoa. Upon Muteba's death in 1920,
Ditende, like some of the other *yikeezy* appointed by the late Mwaant
Yaav, first backed the claims of Mushidi's descendant, Tshipao, but
very soon switched his support to Kaumba. This politic move left him
undisturbed in the enjoyment of his governorship of Mbako, which he
held until his elevation to the throne.[53]

Although Ditende was clearly the administration's favorite, his
election was by no means a routine decision. No fewer than nine
candidates were considered by the council presided over by the
Kanampumb. Three of them were descendants of Naweej Ditende:
Chief Mbako, Chief Lumanga (Mutshaila Ditende), and the late
Kaumba's *Nswaan Mulopw*, Mulolo. Three other candidates were
descended from Muteba, Kaumba's predecessor on the throne; and
three more came from the line of Mwaant Yaav Mbumba. These were
Mwadiata Makaza, a grandson of Mbumba; Lutongo Chief Muyeye, a
son of Kawele; and Gaston Mushidi, a grandson of Mwaant Yaav
Mushidi I whom the Belgians had deposed in 1907.[54] Actually, only

Gaston Mushidi was regarded as a serious alternative against Ditende. The son of *Rukonkesh* Kamina, Gaston Mushidi had been born in 1911 and educated, unlike Ditende, by the Franciscan missionaries.[55] His other major distinction, which later turned out to be of decisive importance, was the fact of being Moïse Tshombe's maternal uncle. Tshombe's father, Joseph Kapenda, the successful merchant, plantation owner and founder of the family fortune, had died earlier that year and the future president of the State of Katanga, his eldest son, had now become head of that influential family. Despite the fact that he had married one of Ditende's daughters, Ruth Matshik, in 1938, Moïse Tshombe and his brothers David, Daniel, and Thomas backed the candidacy of Gaston Mushidi, brother of their mother Kat.

The council strongly inclined toward Mushidi, partly out of an unexpressed sense that the dignity of Mwaant Yaav should rotate among the several competing lineages. But, for years, Chief Mbako had consistently been rated by the administration as one of the best chiefs in all of Katanga, and Belgian officials who attended the council made no secret of their preference. Ditende was elected and the administrator for Kapanga exulted:

> Unlike Kaumba, who only utilized [his prestige and authority] for personal aims, Bako Ditende uses them also for administrative ends. Thus, we do not hesitate to ask for his opinion in all matters relative to the administration of the chefferie and to relay all important directives through his intermediary. For his part, he solicits our advice over all major issues and takes no important initiative without informing us.[56]

Tension ran high at the Musuumb as the feud continued between the two contending factions. Ditende accused Gaston Mushidi of having used poison to cause the death of Kaumba, then tried to have him arrested for poaching. Moïse Tshombe, who had just been made a member of the provincial Conseil de Gouvernement (a toothless sounding-board advisory to the governor) tried to use his influence on behalf of his uncle, but to no avail. Clearly, the administration was not going to let the disgruntled Mushidi undermine the new Mwaant Yaav's position, even if the Tshombe brothers were regarded as models of acculturation. The Lunda council's recommendation that Mushidi be appointed to fill the now vacant post of Chief Mbako was ignored. Finally, in December 1952, after a long series of incidents, Gaston

Mushidi was deported at Ditende's request to the central Katanga town of Mwanza.[57] Meanwhile, at the end of May, Ditende, who had assumed the title of Yaav a Naweej III, had been ritually invested at the Nkalaanyi with all the customary ceremonial. Not so traditional perhaps, but equally symbolic of another form of recognition was the gift of a brand new Ford automobile which he received on that occasion from the three largest private concerns in Katanga — the BCK railroad, the Comité Spécial du Katanga, and the Union Minière.[58]

7

From Tribalism to Ethnicity:
The Growth of Lunda and Cokwe
National Sentiments

Ditende's accession to the dignity of Mwaant Yaav inaugurated an era of unprecedented cordiality in the Lunda court's relationship with the colonial administration. In contrast with the thinly veiled annoyance so often manifested by the Belgian authorities during the last years of Kaumba's reign, Ditende's performance was the object of continuous praise from the moment of his installation: "Conscientious, dedicated, active, loyal" were some of the adjectives that kept cropping up in the reports dispatched from Kapanga during the mid-fifties, and the new Mwaant Yaav consistently received a near-perfect rating of 24 points (out of a possible maximum of 25) after each annual evaluation of his record. In sum, Belgian officials concluded, Ditende is "really on our side," "truly a model chief in whom we can place our entire trust."[1]

Official appreciation for the new Lunda paramount also took a more concrete form. In 1953 Ditende, along with a handful of other "model chiefs," was flown to Belgium for an official visit and, upon his return, he duly convened the population of his capital (as well as the resident Belgian administrator, of course) to hear his admiring relation of all the wondrous things he had seen and learned in the *métropole*. How could official circles fail to be impressed by Ditende's wisdom? Shortly thereafter he was appointed to the advisory Conseil de Gouvernement which had only recently been enlarged to include eight African members (all of them selected, in the words of M. Crawford Young, "presumably on the basis of their suitability from the administration's viewpoint") and to its equally consultative local counterpart, the Conseil de Province.[2] In 1955, during King Baudouin's well-orchestrated first visit to the Congo, Ditende was awarded the privilege of a personal encounter with "Bwana Kitoko."[3]

The Mwaant Yaav's cooperativeness was valuable to the administration for several reasons. The haunting fear of a colonial uprising which had loomed so large after 1944[4] had now almost completely receded, and many Belgians had virtually persuaded themselves that the winds of change might not blow their way for another generation; but quite a few farsighted members of the colonial establishment, whether officials, businessmen, or missionaries, realized that the process of social change was irreversible and that it was only a question of time (how little time, in fact, few of them seemed to guess) before Belgium would have to face the same problems as the other colonial powers. The number of urbanized Africans had grown from slightly over one million in 1940 to 2,162,397 in 1950, and the most dramatic increases had been registered in the largest cities, with the African population of Elisabethville rising from 27,000 in 1940 to 99,000 in 1950 and that of Léopoldville multiplying more than fourfold over the same ten years.[5] During the same period, the number of African wage-earners had increased from 536,055 to 962,009 and the rural exodus showed no signs of slackening. Like the sorcerer's apprentice, the administration now watched with a strange admixture of pride and awe this rising tide which it had strained to initiate a generation earlier and which was now moving of its own momentum, seemingly beyond anyone's control, and with ominous political implications. Efforts made in 1945 to contain this migrating flow by tightening the pass system[6] had been of little avail in the face of expanding opportunities for employment in the urban centers. Increasingly, therefore, the administration's attempts to stabilize the rural population had taken the form of offering the villagers a more attractive alternative to urban employment than the compulsory cultivation system generalized during the depression and the war years. On the basis of limited experiments initiated as early as 1936 by the INEAC (Institut National pour l'Etude Agronomique au Congo Belge) and in 1941-1942 by cotton export companies, the system of *paysannats indigènes* was gradually introduced throughout most of the Congo, beginning in the late forties. Crop rotation and the establishment of a viable balance between food and cash crops were the primary objectives of the *paysannats*. Unlike previous attempts to "educate" the rural population, however, they

were supposed to be largely voluntary. In fact, the cotton companies'
stake in maintaining production levels and the administration's own
traditions of authoritarian paternalism inherited from an earlier era
combined within a short time to reduce many, if not most, of the *paysannats* to no more than a disguised version of the compulsory crop
system.[7]

Whether they were seen as counterweights against the potentially
unmanageable masses of urbanized Africans, as possible catalysts in
the attempt to stabilize the rural population, or merely as useful
auxiliaries to ensure compliance with the administration's latest
agricultural schemes, cooperative chiefs such as the new Mwaant Yaav
were invaluable to the Belgian authorities. Actually, compared with
other regions of the Congo, Lunda offered no serious resistance to the
introduction of cotton. Without ever becoming as productive as some
of the non-Bantu areas of the north or even the neighboring Kaniok
and Luba regions, each of the three territories of Kapanga, Sandoa,
and Dilolo managed to grow an average of between one and two
thousand metric tons of cotton annually. This was not inconsiderable
for an area which had failed for so long to develop any viable
cash-crop economy, and a limited measure of prosperity spread to the
central and northern portions of the Lunda heartland. Thus,
Ditende's authority was not greatly taxed by his efforts on behalf of the
administration's agricultural policies, but when it came to containing
the rural exodus, there was little that the Mwaant Yaav or anyone else
could do.

The combined tribal population of Kapanga and Sandoa territories,
which had exceeded 110,000 in 1939 (see Table 13) numbered 104,973
ten years later and 108,079 in 1957, just as the effects of the recession
suffered on the Copperbelt were beginning to drive men back to the
rural areas. By contrast, the tribal population of the southern territory
of Dilolo[8], which had declined steadily until the end of the war, was
now on the increase and moved ahead of its prewar level after 1952
(see Table 14). The villagers of Dilolo territory not only took to the
planting of cotton without much reluctance (in 1951 the Belgian
administrator reported that even farmers not subject to compulsory
cultivation were asking him for cotton seeds)[9] but were also able to
market a significant proportion of their food crops locally to the
growing salaried population. While store clerks and railroad crews
continued to represent a substantial portion of the wage-earning

Table 14. Kapanga, Sandoa, and Dilolo: Tribal and non-tribal
populations, by territoire, 1947-1958.

Year	Kapanga		Sandoa		Dilolo	
	Tribal	Non-Tr.	Tribal	Non-Tr.	Tribal	Non-Tr.
1947	43,779	791	63,701	701	70,216	7,610
1948	42,842	1,601	62,737	(not avail- able)	69,404	6,209
1949	43,266	1,794	61,707	3,217	73,479	6,359
1950	44,053	2,462	61,161	2,952	74,233	5,763
1951	45,710	3,394	60,153	3,436	74,525	6,198
1952	45,685	3,805	60,364	4,417	75,610	8,069
1953	45,967	3,909	59,936	3,655	78,630	7,554
1954	44,019	4,823	61,155	3,025	79,085	10,605
1955	43,918	3,644	62,200	2,934	81,074	9,936
1956	43,581	4,124	63,561	2,920	80,830	11,652
1957	44,224	4,305	63,855	3,011	85,468	10,548
1958	43,634	4,133	70,438	1,636	91,242	10,476

Source: Territoires de Kapanga, Sandoa, and Dilolo; Rapports
AIMO 1947-1958.

population of Dilolo territory, the development of the Kisenge manganese mine exploited by the BCK railroad company[10] offered a new
source of employment and created a small-scale local replica of the
Copperbelt company towns. The proportion of Africans living outside
the tribal structure increased dramatically over the whole of Lunda,
from little more than 3 percent in 1944 (and less than 2 percent in
1939) to 9.1 percent in 1954. Most of this nontraditional population
naturally collected around the commercial and administrative centers
of Sandoa, Dilolo, Kapanga, Malonga, and Kasaji, in the Kisenge
mining area, or along the railroad, but the tribal population also felt
the impact of these economic poles in the sense that those chefferies
located in their vicinity registered a normal demographic increase,
indicating that, in their case at least, the availability of economic and
social opportunities had partly stemmed the drain of rural migration.
Thus, the traditional communities registering the strongest population
increase between 1949 and 1958 were the chefferies of Mbako and
Lumanga (near Sandoa), Dumba, Saluseke, Tshisenge, and Kandala
(near Dilolo), Tshisangama, Samujima, Sakayongo, and Tshanika

Table 15. Kapanga, Sandoa, and Dilolo: tribal population by
native administration unit, 1949-1958.

Native administration unit	1949	1952	1958
Mwata Yamvo (L)[a]	43,266	45,685	43,634
Tshibamba (L)	4,898	5,537	5,203
Muteba (L)	4,752	4,867	5,394
Mbako (L)	10,747	11,724	13,519
Lumanga (L)	12,392	13,908	17,178
Kayembe Mukulu (L)	7,739	7,678	7,382
Tshipao (L)	5,954	5,936	6,496
Dumba (L)	8,709	9,591	11,713
Muyeye (L)	7,170	7,024	7,523
Tshisangama (L)	8,188	8,562	11,004
Kazembe (L) ⎫ S. Mutanda[b]	7,191	7,243	⎫ 11,236
Mukonkoto (L) ⎭	1,950	1,817	⎭
Sakayongo (L) ⎫	2,784	2,559	⎫
Samujima (D) ⎬ S. Lulua-Lukoshi[b]	3,033	2,424	⎬ 12,035
Tshanika (D) ⎭	4,809	4,569	⎭
Saluseke (D)	7,431	7,663	10,359
Samutoma (C)	10,754	10,616	11,706
Sakundundu (C)	4,471	4,605	4,678
Tshisenge (C)	6,754	7,689	9,430
Kandala (C)	6,539	7,659	8,183
Katende Muteba (A) ⎫	1,731	1,343	⎫
Katende Jean (A) ⎪ S. Luena[b]	2,861	2,470	⎪
Katende Tshipoy (A) ⎬	2,641	3,025	⎬ 9,759
Tshilemo (A) ⎭	2,088	1,972	⎭

Source: Territoires de Kapanga, Sandoa, and Dilolo; Rapports
AIMO 1949, 1952, and 1958.

[a]Dominant ethnic group: L = Lunda, C = Cokwe, D = Ndembu,
A = Alwena.

[b]These *secteurs* were organized in 1953.

(near Kisenge and Malonga). The tribal population of nearly every
other chefferie in Lunda (including the Mwaant Yaav's) stagnated or
declined (see Table 15).

In such a fluid situation, the Mwaant Yaav's prestige could be of
considerable service to the administration, and we have seen that

Belgian officials showered Ditende with every form of recognition open to an African under the colonial system. The administrator for Kapanga wrote:

> [We] continue to take advantage of every possible opportunity to enhance the Mwata Yamvo's prestige and to reinforce his authority. A place of honor is always reserved for him during all ceremonies taking place in the territory; whenever notables meet in council, they are always reminded that the scrap of authority they wield was was given to them by the Mwata Yamvo, and so on.[11]

But if the administration was prepared to honor Ditende in its own self-serving way, the Mwaant Yaav for his part never lost sight of his traditional position and did not hesitate to draw, when needed, on the credit he had accumulated with the colonial administration. Like his two predecessors — but with all the finesse acquired through a lifetime of dealing with the Europeans — Ditende sought to consolidate his own personal position over a court traditionally given to intrigue and to restore as much as possible his authority over the scattered remnants of the empire. Overlapping these two sets of preoccupations, and linking them to some extent, was Ditende's calculated policy of advancing the political fortunes of the members of his immediate family, preferably at the expense of Mbumba's descendants.

The new Mwaant Yaav's obvious influence with the Belgian authorities may well have stood him in good stead when it came to establishing his ascendancy over the members of the Lunda aristocracy, many of whom were still being courted more or less overtly by the Tshombe family. The banishment of Gaston Mushidi, ordered by the colonial administration on Ditende's suggestion, was followed in July 1954 by that of one of the late Kaumba's retainers, the Mukakatoto, also for refusing to bow to the new paramount's authority.[12] The following year, with the administration's tacit concurrence, the Mwaant Yaav dismissed one of the senior title-holders, the Kanampumb, and replaced him with one of his own appointees.[13] At the same time, the attrition of age resulted in the gradual elimination of members of the old order: thus Kaumba's *Nswaan Mulopw*, Mulolo, died in July 1953 and was replaced by one of Ditende's followers.

While bolstering his own authority over the court, the new Mwaant Yaav also managed to defend his capital against the modernizing zeal of the colonial administration. Ever since the end of the war, but particularly during the last years of Kaumba's reign, the Belgian authorities' obsessive concern with security and productivity had come to focus on the Musuumb. The tiny capital, with a population of 3,382 at the end of 1950, could hardly be viewed as a threat to law and order, but it was sufficiently atypical to constitute a minor irritant. Furthermore, "traditional" though it was, it had an unusually large number of *évolués* (sons of courtiers for the most part) and, more vexing still, it performed no economic function that the colonizer could recognize as legitimate. Thus, during the late forties, Belgian officials gradually persuaded themselves that the Musuumb was badly in need of a *dégorgement* ("disgorging," or "unclogging": the term sounds as peculiar in French as it does in English), and, when the *paysannats indigènes* grew in fashion, they felt that they had hit upon the ideal solution: the "excess" population of the Lunda capital would simply have to be resettled at some distance from the Musuumb. As soon as he felt that he could speak with sufficient authority, however, Ditende strongly objected to the resettlement scheme. "The aim," declared the Belgian administrator, "is to leave at Musumba only the Mwata Yamvo and his notables, with a few close relatives."[14] To which the Mwaant Yaav objected:

> Well, I have to say that it would not be a good thing if the people left Musumba, and that for several reasons. First, Musumba is the capital of the Lunda empire. There is only one Musumba. . . . And then, the Mwata Yamvo has the right to demand of anyone that he should come and reside at Musumba and that he should remain in his immediate entourage . . . . And then, Musumba is the administrative center. All *lukano*-bearing chiefs come from Musumba. Traditionally, a *lukano*-bearing chief will send to Musumba one of his potential successors . . . . And then, you have the senior dignitaries; they must have their men around them . . . . And then the Mwata gives orders to his dignitaries; how do you expect them to carry them out if they have no people around them? . . . First, it would be illogical for me to strip my capital of its rank and of its people and then I will be faced with massive opposition from all the dignitaries. The people leaving Musumba—that would not be a good thing.[15]

In fact, 88 families were resettled in the periphery of the royal village in hamlets controlled by court dignitaries, but the Belgian authorities soon realized that the population of the Musuumb was predominantly composed of court officials, tradesmen, artisans, etc., and that only a minority of inhabitants were farmers, suitable for resettlement on a *paysannat*. By the end of 1953, colonial officials had come to accept most of the Mwaant Yaav's views and concluded that any further reduction in the population of the Musuumb could be achieved only through coercive measures which would be "impolitic" and "harmful to the Mwata Yamvo's prestige," as it would reduce the capital to the size of "a mere headman's village."[16] As it was, the Musuumb had lost over one thousand of its inhabitants, but Ditende had managed to arrest the *dégorgement* process and the population loss was eventually more than recouped after the desultory Europeans had turned their attention to other issues.[17]

Ditende lost no time in attempting to assert his authority over the local Lunda chiefs. The fact that his father had been a Cokwe, or that his accession had been assisted by the colonial administration—like that of Muteba and Kaumba before him—apparently caused no significant revulsion against him. Every major *yikeezy* and even a number of emissaries from Lunda chiefdoms in Northern Rhodesia and Angola had traveled to the Musuumb to pay homage to the new Mwaant Yaav. At the same time, however, having himself been a *yikeezy* for many years, Ditende knew that local chiefs were subject to the permanent temptation to act autonomously and to consolidate themselves and their kin in their respective bailiwicks. The continued grudge of the Tshombe family was a potentially deleterious ferment against which the new Mwaant Yaav wanted to protect himself. Like his predecessors, therefore, Ditende sought to reinforce his control by lending his prestige to those chiefs he judged to be loyal (such as Tshisangama), by undercutting known or suspected opponents and, naturally, by appointing members of his family to vacant governorships. His own former position of Chief Mbako, which the Lunda council had suggested should go to Gaston Mushidi as a sort of consolation prize, was awarded to Ditende's son, Paul. In 1956, another son, David Dibwe, was peremptorily appointed over the objections of local headmen, to succeed to the title of Chief Muyeye. This office, it will be recalled, had been held by Lutongo, a grandson of Mbumba, who had been regarded, next to Gaston Mushidi, as the strongest candidate against Ditende in the 1951 election.[18] Although *yikeezy* titles are normally appointive, Lutongo's son had hoped to succeed his father and had apparently won the support of a majority of local notables, but the Mwaant Yaav maintained that governorships were not hereditary as of right and insisted on exercising his discretionary choice. The colonial administration backed up Ditende who, having scored one more point against the descendants of Mbumba, proceeded to show his magnanimity by appointing a younger son of the late chief as *Nswaan Mulopw* to the new Muyeye.[19] The following year, the Mwaant Yaav was even invited to nominate a chief for a community in Northern Rhodesia. On that occasion, he traveled to the neighboring British territory and dispatched one of his own daughters as chief for Chavuma.[20]

None of the Mwaant Yaav's actions during the early years of his reign would seem to have departed fundamentally from the dynastic policies followed by his predecessors. In particular, apart from his reference to the Musuumb's unique position (which could be regarded as a mere statement of fact), there was little in Ditende's early political moves to suggest a deliberate effort on his part to promote the mystique of a restored Lunda empire, the promise (or threat) of which was to acquire such conflictual significance only a short time later. As was the case with the proposed revival of the Kongo kingdom (admittedly in a rather different context), the somewhat nebulous idea of an imperial "restoration" seems to have originated with the younger generation, specifically with men who belonged by their birth in the ranks of the Lunda establishment but had received a measure of Western education, had followed nontraditional careers (usually in white-collar occupations), and had spent all or part of their adult lives outside their traditional communities.[21] At the same time, the linguistic and cultural components which had become so prominent in the growth of Kongo nationalism and had provided ABAKO with its initial impetus were almost totally lacking in Lunda, for the simple reason that political, not cultural integration had been the hallmark of the Lunda system. The KiCokwe language was more widely spoken in Sandoa and Dilolo territories than was Iruund, the distinctive idiom of the northern Lunda, and any effort to promote the latter as a "national" language might well have exposed the numerical weakness of the true Aruund. Only history and institutions, rather than strict ethnicity, could serve as foundations for Lunda "national" sentiment and no one could better incarnate these twin determinants than the Mwaant Yaav. Thus, while Kongo nationalism could (and did) flourish independently of the rather pitiful nonentities enthroned at São Salvador, Lunda nationalism (comparable on that score to its Ganda or Ashanti counterparts) could not exist without the Mwaant Yaav. Cokwe nationalism, by contrast, was able to combine the galvanizing effect of historical memories (particularly those relating to their disruption of the Lunda empire) with a grass-roots appeal to the sense of relative deprivation felt by many Cokwe toward their more prestigious neighbors.[22]

Cokwe nationalism originated from many different sources, reflecting the lack of a unified political tradition as well as the wide dispersion and high mobility of this tribe. In Katanga itself, it took the form of demands for equal recognition on the part of Cokwe chiefs and of the organization of voluntary ethnic organizations by urbanized members of the tribe, both efforts converging in the attempt to mobilize the support of the thousands of Cokwe villagers who lived under the authority of Lunda chiefs.

The demands for equal treatment on the part of Cokwe chiefs and their insistence on repudiating any form of subordination toward the Mwaant Yaav were of course nothing new. They had been responsible at least in part for inaugurating a policy of ethnic autonomy in Lunda during the early 1920s, and for its being maintained during the second half of that same decade despite the objections of several field officials. They had led also, after 1930, to the lapsing of any further payment of the disguised tribute to the Musuumb which the Belgian authorities had originally tried to preserve under the euphemistic name of *redevance*. In the case of the largest Cokwe chefferie, Samutoma (initially known as Mwatshisenge), the search for political autonomy had been complicated ever since the 1920s by the fruitless yet persistent claims of successive chiefs to a position of paramountcy over all Cokwe, paralleling the Mwaant Yaav's status among the Lunda.

Although the Belgian authorities had prevented its continued use, the chiefs of the Cokwe lands lying to the west of Sandoa had always maintained their pretensions to the prestigious title of Mwatshisenge, and minor incidents with neighboring Lunda chiefs over questions of status had been recurrent through the 1940s, notably upon the installation of a new Samutoma in 1949.[23] Such claims to paramountcy over the whole Cokwe nation (or at least over its Katanga branch) were far from being accepted at face value by the three other Cokwe chiefs, none of whom, however, was in any position to offer counterclaims of their own.[24] The prospect of a Cokwe national revival led by a chief was further complicated by the fact that only a minority of the Cokwe villagers residing in Katanga actually lived under the authority of a chief of their own tribe; the rest were settled on lands ruled over by alien chiefs (most of them Lunda), sometimes as Cokwe enclaves (such as Sapindji, located in Lumanga's chefferie, between Sakundundu and Samutoma), or in scattered villages where a good deal of intermarriage had taken place. The Cokwe also represented an overwhelming majority of the immigrants who had continued to flow into Katanga from Angola throughout the colonial period, further adding to their numbers in the southern half of the Lunda heartland.[25] Only a fraction of those immigrants actually settled on lands ruled by Cokwe chiefs, however, because of the small size and overcrowded conditions of the Cokwe chefferies, particularly in the south.

The areas of Lunda transferred under the authority of Cokwe chiefs by the Kapanga convention of 1923 had been too small, even then, to accommodate the entire Cokwe population of the region, and the notion of regrouping all the members of that tribe in the lands controlled by their own chiefs had accordingly never been regarded as a serious possibility. Together, the four Cokwe chefferies represented approximately one-sixth of the joint area of Sandoa and Dilolo territories, and accounted for over one-fifth of their combined population. But while the density of land occupation was about average in Samutoma and sparser than average in Sakundundu, the two southern chefferies of Tshisenge and Kandala and a population density of 56 inhabitants per square mile—seven times the average density for Dilolo territory as a whole. With 19 percent of the tribal population of that territory, they represented together less than 3 percent of its land area. In a study prepared in 1957, the administration estimated that even under the best circumstances (i.e., if scattered holdings could be regrouped, etc.) there would not be a single square foot of arable land available in Tshisenge by the end of 1961.[26] Ever since the early 1950s, therefore, Belgian officials had been studying ways of relocating the excess population of the two southern Cokwe chefferies in neighboring areas. Several regions were considered and their respective chiefs approached, but all of them expressed strong reservations about welcoming the aggressive, fast-expanding Cokwe. The Lunda chiefs, for their part, made it clear that while they might be willing to settle individual Cokwe families, they would not accept any of their headmen.[27]

But the full dimensions of the Cokwe land problem appear even more starkly when it is considered that although only one-sixth of the combined areas of Sandoa and Dilolo territories was ruled by Cokwe

chiefs, the tribe itself represented a clear majority of the total population of these two divisions. How large a majority? This is almost
impossible to estimate, if only because many Cokwe living under the
authority of Lunda chiefs were treated as Lunda by the administration
and would often identify themselves as such, whether out of personal
choice or because of a land-centered sense of political allegiance. Estimates made at various times between 1923 and 1960 by local Belgian
officials had recorded Cokwe majorities in the Lunda chefferies of
Mbako, Muteba, Mwene Ulamba, Kayembe Mukulu, Tshipao,
Tshisangama, Mukonkoto, and Dumba, and important minorities virtually everywhere else. The Belgian administrator for Sandoa estimated
in 1958 that the Cokwe represented 65 percent of the population in
Dilolo territory,[28] while the district commissioner for Lualaba, analyz-
ing the political situation in early 1960 without any particular sympathy for the Cokwe, placed their strength in Sandoa territory at 64
percent of the population.[29]

As the Congo moved into the realm of participatory politics without
the cushioning effect of a qualified franchise (different in that from
both British and French territories), the political potential of the
Cokwe vote became a crucial factor on the Katanga scene. Could the
entire Cokwe tribe be mobilized (if only for electoral purposes) behind a
single organization? And, if so, would it throw in its weight with other
"authentic Katangese" groups into a regional alliance, or attempt to
assert itself at the expense of its neighbors? The answer to these questions clearly lay beyond the grasp of any one of the four Cokwe
chiefs—even of Samutoma, the more politically minded of the lot—
and depended in large part upon the activities of the "detribalized"
Cokwe communities.

Voluntary ethnic associations had made their official appearance
rather late in Katanga. The first such organization of any significance,
the Association des Baluba Centraux et Kasai au Katanga (later
renamed Fegebaceka) had been authorized in July 1955 and consisted,
typically, of the more aggressive and worldly-wise Luba immigrants
from Kasai. Three months later, the first Cokwe organization, the
Association Tshiyande Tshokwe, gained official recognition in Kolwezi.[30] Other Cokwe associations were successively organized during
1956 and 1957 in the western Katanga copperbelt centers of Jadotville,
Shinkolobwe, and Lubudi, where the concentration of Cokwe laborers
was the greatest.[31]

Meanwhile, however, a more ambitious movement aiming at the organization of the whole Cokwe nation had been initiated in Elisabethville. The Association des Tshokwe du Congo, de l'Angola et de la Rhodésie (Atcar) was founded in November 1956 by Ambroise Herman Muhunga, a nephew of Mutunda, Chief Samutoma. Muhunga had resided alternately in Elisabethville and Dilolo for over twenty years and had made his living variously as a petty trader, a white-collar worker for several European firms, and most recently as a small independent businessman of indeterminate means.[32]

The exact circumstances and date of Atcar's foundation are far from clear. In a letter written in January 1957 to justify his fund-raising activities, Muhunga merely refers to the association as having been founded "in 1956," but its by-laws were apparently not registered with the authorities until July 1, 1957.[33] A possible clue to the uncertainties surrounding the birth of Atcar may lie in the interaction between Cokwe communities in Katanga and Northern Rhodesia during this period. In October 1956, after several months of planning, a Cokwe self-help association known as Ukwashi wa Chokwe was founded at Mufulira, on the Northern Rhodesia Copperbelt, under the chairmanship of one John K. D. Kajila. According to leaders of that organization, Muhunga had visited several Northern Rhodesia mining towns in August 1956, "selling jackets and women's clothes" and had "stolen the idea for his movement" from his fellow tribesmen at Mufulira.[34]

It may be that the lack of precision regarding the exact date of the founding of Atcar resulted from a deliberate effort on Muhunga's part to obfuscate this issue, but it appears unlikely that he would have needed the example of the Mufulira groups to conceive of his movement: after all, Cokwe mutual associations had been founded in Kolwezi and Jadotville prior to the summer of 1956, and the idea of organizing the Cokwe was clearly in the air. We do know, at any rate, that in early January 1957 Muhunga was engaged in an aggressive fund-raising campaign in Dilolo, where he had maintained many contacts,[35] and that by the time Muhunga and the Northern Rhodesia group entered into formal contact in the spring of 1958, Atcar was undoubtedly the more dynamic—and also the more politically-minded—of the two organizations.[36]

The efforts made by the urban Cokwe associations to organize the rural Cokwe gathered momentum from the beginning of the year

1957. Spreading from local centers of salaried employment such as Dilolo, Sandoa, Kisenge, Kasaji, Mutshatsha, etc., they rapidly infiltrated the rural hinterland, without much apparent coordination. In the communities ruled by Cokwe chiefs, they found a relatively easy terrain for their expansion. Chief Tshisenge (of Dilolo) gave the movement his wholehearted endorsement, perhaps to enhance his questionable legitimacy, but also as a means of articulating the long-standing grievances of his people regarding their acute shortage of arable land. Tshisenge's support for the movement, on the other hand, was viewed by his neighbor, Chief Kandala, as a threat to his own position. During the early 1950s, Kandala had resisted the Belgian administration's recurrent idea of amalgamating the two adjacent Cokwe chefferies because of his not unjustified apprehensions that his neighbor, not he, would emerge as chief of the proposed secteur. Kandala, therefore, did nothing to encourage the organization of a Cokwe association in his bailiwick, and his subjects (whose lands, incidentally, were reasonably adequate for their needs) were accordingly far less responsive to the whole idea.[37]

At Samutoma, the ruling family's persistent claims to paramountcy over the whole Katanga branch of the Cokwe nation offered both promises and problems. The reigning Samutoma (Muhunga's maternal uncle) had been trying for years, like his predecessors, to persuade a rather skeptical administration that he should be regarded as a spokesman for the whole Cokwe tribe. When he died, in February 1957, his succession immediately became a delicate political issue, and a test of strength for the urbanized elements. Muhunga himself, though technically eligible, probably felt that Elisabethville was a more suitable theater for his political activities (not to mention his business interests), but he and other militant Cokwe from Jadotville, from Kolwezi, etc., were intent upon securing the election of a man who would be sympathetic to their cause. Their attempts were frustrated, however, when the council (firmly backed by the administration) selected the late chief's *Nswaan Mulopw*, Thomas Samazembe, to succeed to the title of Samutoma. Yet Muhunga's extraction, as well as his advocacy of a Cokwe national revival in a community which prided itself upon the historical seniority of its chiefs, assured him of a continued leverage at Samutoma. the *évolués* and the traditionalists found a common ground in their shared hostility toward the Lunda. The council's first recommendation to the new Samutoma was to refrain from any action that might imply any form of allegiance or subordination toward the Mwaant Yaav.[38] Samutoma hardly needed any encouragement along these lines and upon his formal confirmation as chief, in May 1958, he insisted upon adopting the historically prestigious title of Mwatshi-senge. He and Muhunga developed a working relationship (tempered in part, it would seem, by a mutual suspicion of each other's ambitions) which lasted until the end of 1959. Thus, while the influence of the *évolués* was being contained, the mounting tension between the Cokwe and Lunda was in no way assuaged by the outcome of the Samutoma succession dispute.

In the lands ruled by Lunda and other non-Cokwe chiefs, the Cokwe awakening was of course viewed as a distinct threat. The administrator for Dilolo noted ominously:

The Tshokwe, who are currently scattered over the Lunda lands, do not realize that they represent a large majority of the population in this territory. Led and advised by the founders of [Atcar], they will become aware of their strength and use it toward political ends to supplant the Lunda chiefs and to take back from them the lands they claim to have conquered before the Europeans' arrival. In fact, the chiefs understood at once the danger which the Tshokwe association represented for them.[39]

In addition to explicit attempts to discourage their Cokwe subjects from joining Atcar (often by intimidation), the strategy of the Lunda chiefs was, in most cases, to promote territorial, rather than ethnic, affiliations by sponsoring *associations de ressortissants* open to all residents of their chefferies, irrespective of their tribal identity. The Groupement des Associations Mutuelles de l'Empire Lunda (Gassomel), later to become one of the principal founding groups of the Conakat, was subsequently organized by Tshombe to channel the political potential of these local associations.

Meanwhile, the Mwaant Yaav was beginning to feel the indirect effects of the Cokwe awakening and was being pressed from various quarters to adopt a firm posture in response to the militancy of the Cokwe. Local Lunda chiefs seeking support against the incipient rebellion of their Cokwe subjects; educated or urbanized sons of the traditional elite, aware of the changing political climate and of the uses of organized strength; members of the court cultivating imperial fantasies — all urged the Mwaant Yaav to follow a hard line and to assert his prerogatives or, preferably, to lend his prestige to their efforts to mount a Lunda counter-offensive, and when Ditende failed to respond with sufficient vigor (in their eyes) a period of tension (involving, inter alia, a deluge of anonymous letters denouncing the Lunda paramount) ensued at the Musuumb. For the ageing Mwaant Yaav, such a campaign could be attributed only to the intrigues of the Tshombe family. Even court dignitaries, he felt, were being insidiously turned against him. Thus, when the *Nswaan Muruund* Kamina accepted a role from Moïse Tshombe between Sandoa and the Musuumb, he immediately concluded that she must have switched her allegiance to his rivals, and forced her to relinquish her position.[40]

In fact, though highly conscious of his rank, Ditende apparently felt (possibly because of his mixed ancestry) that the internal logic of his office and the political traditions of the Lunda state prevented him from even admitting the existence of a Cokwe problem—except perhaps to denounce it as a fabrication of outside agitators. The image (not entirely fictional) of peaceful coexistence and integration among the different ethnic groups living under the authority of Lunda chiefs was to be maintained at all costs. There would be no Aruund party that could indirectly justify Atcar's pan-Cokwe ideology—especially if, as seemed likely, it was going to be dominated by the Tshombe faction.[41] Even Ditende's half-hearted subsequent endorsement of Conakat could be rationalized in terms of that party's claim to speak for all "true Katangese." Whatever the merits of that position, however, it was not acceptable to the Cokwe nationalists and the result was to drive the Mwaant Yaav gradually closer to the Lunda hard-liners, if only to avoid isolation. Thus, his public expressions of support for Mbalandji, Chief Tshisangama, whose predominantly Cokwe subjects were being actively organized by Atcar, clearly did nothing to moderate that chief's notorious persecution of all Cokwe activists.

The Belgian administration's reactions toward the attempts to organize the Cokwe ranged from suspicion to outright hostility, not because of any particular partiality against that tribe, but simply because of the movement's threat to disrupt the status quo—in the form of what a number of officials variously described as the "equilibrium," the "mutual understanding," or even the "symbiosis" which supposedly prevailed between Lunda and Cokwe.[42] The authorities' strongest objections were directed against Muhunga's efforts to consolidate all local Cokwe associations into a single movement. In November 1957, Muhunga launched a three-week organizing circuit which took him to Jadotville, Kamina, Sandoa, Dilolo, and Kasaji[43] —under close watch of the security police which, in one report, characterized him as "rather unscrupulous and very xenophobic." The administration's strategy, at that stage, was to insulate the rural areas from the influence of the Cokwe activists and, without "victimizing the Tshokwe," to discourage the emergence of a unified association covering the whole of Katanga on any but a de facto basis.[44] Statutory rules governing native associations[45] provided a legal excuse by specifying that in nontraditional communities such associations were to be authorized by the district commissioner (and thus implicitly limited to the area under his jurisdiction), while in traditional milieux they were placed under the exclusive authority of the native chiefs. A literal enforcement of these norms could see to it that no Cokwe association would be permitted to operate in any community ruled by a Lunda chief. This corresponded precisely to the recommendations of local Belgian administrators who suggested that the activity of these associations should be restricted to the townships and to the Cokwe chefferies.[46]

Muhunga's November tour of the southwest pulled the rug from under this strategy by rekindling the ideal of Cokwe solidarity throughout Katanga, and although there were no serious incidents during this episode, several exasperated Lunda chiefs came to the Belgian authorities to demand that Muhunga be enjoined from returning to the area. While this particular demand was not met until some time later, the chiefs were assured that Atcar had not been authorized to organize on a provincial scope and that they still remained free to control the activities of any African association in their respective chefferies.

Yet, despite the commotion created throughout Lunda by the first stirrings of Cokwe nationalism, Muhunga and his followers were still far from representing a formidable force. This was indirectly demonstrated a few weeks after Muhunga's campaign swing through southern Lunda when, for the first time in the Congo's history, elections were held in three major cities—Léopoldville and the two largest urban centers in Katanga: Elisabethville and Jadotville. Even though this first step in the direction of urban self-government was explicitly presented as experimental, restricted to three cities,[47] and the elections themselves officially termed "consultations," they provided the first real test of organized political behavior in the Congo and set patterns of ethnic voter alignment that were to weigh heavily on the future of the entire country.

The single-member ward system selected for this election, combined with the administration's refusal to authorize African political parties, meant that the best disciplined ethnic associations were in a position to maximize their influence. Thus, in Léopoldville, the Abako was able to capture 133 of 170 council seats with only 46 percent of the total vote. A similar situation prevailed in Elisabethville and Jadotville, where all African burgomasters came from ethnic groups based outside Katanga and where the political scene was dominated by the Luba immigrants from Kasai.[48]

Neither the Lunda, who represented a mere 6.3 percent of the adult male African population in Elisabethville and 7.9 percent in Jadotville-Kikula, nor the Cokwe (3.33 percent in Elisabethville, 4.7 in Jadotville)[49] were numerically in a position to play a determinant role in these first elections, but the analysis of voting patterns suggests interesting conclusions as to the degree of cohesiveness of the two groups. In Jadotville, which is located closer to the Lunda heartland than Elisabethville and has accordingly a higher proportion of its residents originating from that area, only the African township of Kikula, representing 59 percent of the total African adult male population (along with the European residential section, of course) was permitted to elect a *conseil communal*.[50] The ethnic breakdown of eligible (i.e., adult male Congolese) voters in Kikula is shown in Table 16. Out of the 10,707 possible voters, 1,448 were declared ineligible (mostly because of residence requirements) and another 1,649 failed to register. The number of valid ballots actually cast on election day (December 22, 1957) was 5,996. The breakdown of these votes in terms of the ethnic identity of the 96 candidates who competed for the seventeen council seats appears in Table 17.

From a comparison of the data offered in these two tables, it readily appears that ethnic solidarity did not operate along any simple pattern of predictability. The magnetic pull of certain candidates may be explained in part by individual factors but, taken collectively, it tends to confirm some of the empirical notations often made by observers (outside as well as inside Zaïre) concerning the greater "prestige" of certain ethnic groups. In Kikula this "prestige" factor apparently worked most dramatically in favor of the Lunda (22.1 percent of the vote with only 8.6 percent of the electorate), of the Kasai BaLuba (35.1 percent of the vote with 22.1 percent of the electorate) and—to a more uncertain degree—of the Songye (Kabinda) and Tetela groups.

Table 16. Kikula Township: ethnic distribution of eligible voters (December 1957).

Ethnic group	Number of potential voters	%
BaLuba (Kasai)	2,366	22.1
BaLuba (Katanga)	1,916	17.9
BaSanga	944	8.8
Lunda	916	8.6
Cokwe	549	5.1
BaLomotwa	381	3.6
Ndembu	340	3.2
BaZela	175	1.6
BaTabwa	167	1.6
Kabinda	133	1.2
BaTetela	67	0.6
Others	2,753	25.7
Total	10,707	100.0

Source: Adapted from A. Binamé, "La consultation du 22 décembre 1957 dans la Commune de Kikula" (Ville de Jadotville, 30 April 1958), pp. 35-36.

By contrast, the failure of the Sanga (3.8 percent of the vote with 8.8 percent of the electorate) and Cokwe (2.8 percent of the vote with 5.1 percent of the electorate) to rally any substantial portion of the vote of their fellow-tribesmen would seem to indicate a low degree of tribal identity and cohesiveness, or, interchangeably, a lack of mobilizing organization. Organization also paid off in other ways: thus, the large number of council seats captured by the Katanga BaLuba with only 15.5 percent of the vote is undoubtedly due in large part to the fact that their candidates competed with one another in only three of Kikula's seventeen wards. Similarly, Lunda candidates were in competition in only four wards, while the Kasai BaLuba, far from being as well organized as they were reputed to be, entered no fewer than forty-one candidates who competed against one another in twelve of the seventeen wards.[51] The contrast between the Lunda success and the Cokwe failure in the Kikula election is all the more striking in view of the fact that an undetermined number of the voters listed as Lunda by

Table 17. Kikula Township: ethnic distribution of votes cast in
the December 1957 election.

Ethnic group of candidates	Accumulated no. of valid votes cast for candidates of same ethnic group	%	No. of seats
BaLuba (Kasai)	2,106	35.1	4
BaLuba (Katanga)	930	15.5	5
BaSanga	229	3.8	1
Lunda	1,329	22.1	3
Cokwe	166	2.8	
BaLomotwa	47	.8	
Ndembu	60	1.0	
BaZela	76	1.3	1
BaTabwa	82	1.4	
Kabinda	384	6.8	1
BaTetela	147	2.5	1
Others	440	7.3	1
Total	5,996	100.0	17

Source: Adapted from A. Binamé, "La consultation du 22 décembre 1957," pp. 36-38.

the administration were probably Cokwe originating from Lunda chefferies. But then again, the fact that most of them apparently did not react against this identification and presumably ended up voting for Lunda candidates indirectly confirmed the Lunda aristocracy's claim as to the degree of integration between the two groups — or (from the Atcar's viewpoint) how much work remained to be done to bring about a Cokwe *prise de conscience.*

Whether or not they realized the fragility of the Cokwe movement, the administration and the Lunda chiefs spent the next few months attempting to roll back its activities in the southwest, or, as one official put it, "to annihilate Ambroise Muhunga's nefarious influence."[52] What this implied on the part of the chiefs can only be surmised. One cranky old settler from Kasaji, acting as Atcar's self-appointed champion, deluged the provincial authorities with excited letters in which he described the various forms of intimidation to which the local Cokwe organizers were allegedly subjected on the part of Chief Tshisangama

, (who was indeed described in one report as being "often intransigent" in his dealings with his Cokwe subjects), but we have no information as to what went on in other areas.[53] From the administration viewpoint, defusing the Cokwe time bomb involved several related policies: working on the Cokwe chiefs to persuade them to curb the militancy of local Cokwe associations, but at the same time attempting to meet some of their material grievances, such as the Tshisenge land problem. Chiefs Sakundundu and Kandala posed no great problems, as we have seen. Chief Tshisenge, who had shown considerable sympathy for Atcar, was eventually persuaded to dissociate himself from the movement.[54] Chief Samutoma was harder to convince. After the Mwaant Yaav had traveled to Elisabethville in March 1958 to complain about Atcar's activities, the district commissioner for Upper Lomami convened a general palaver of chiefs from Sandoa territory to bring about a reconciliation. The chiefs reportedly agreed that ethnic associations "served no useful purpose and only fostered civil discord,"[55] but this did not prevent Samutoma from subsequently accompanying Muhunga on the campaign trail, nor from assuming the title of Mwatshisenge when he was officially invested two months later.[56]

Meanwhile, the administration dragged its feet as much as it could on the issue of whether Atcar should be authorized to operate as a single association throughout Katanga. The unofficial recommendation of de facto tolerance formulated in November 1957 by the governor was no longer adequate, whether for those who wanted to curb the Cokwe movement[57] or for Muhunga himself, who wanted a clean-cut ruling from the administration. During the spring of 1958, he again petitioned the governor to authorize a single association instead of having distinct local organizations as the administration wanted. At the same time, he also asked for permission to tour what he insisted on regarding as local branches of the association in the southwest. The Belgian authorities stalled Muhunga, first with the red herring of Atcar's denomination which, they claimed, was inacceptable since it appeared to imply activities outside the Congo itself.[58] Then, when Muhunga indicated that he would be willing to change the name of his movement, they suggested as a compromise that recognition might be granted at the territoire level. This was not acceptable to Muhunga and so his two requests were eventually turned down, in accordance with the recommendations of a highly unfavorable position-paper drafted by the Service of Native Affairs and Labor (AIMO).[59]

Much of the Katanga administration's policy toward Atcar at that stage was clearly (and consciously) inspired by precedents drawn from other regions of the Congo where ethnic associations (notably Abako and, to a lesser extent, the "Lulua-Frères") had adopted a militant tone and tended, as a result, to be regarded as subversive — a view that persisted in some circles, at least with respect to Abako, until the spring of 1960. In the meantime, however, Belgian policy was changing faster than ever before. In October and November 1958, a Belgian parliamentary delegation (the so-called *Groupe de Travail*) traveled throughout the Congo on a fact-finding mission, preliminary to the drafting by the cabinet of a comprehensive policy declaration regarding the country's political future.[60] This development prompted several politically-minded groups to give their efforts a more structured appearance. On October 10, 1958, the Mouvement National Congolais (MNC) came into formal existence under the leadership of Patrice Lumumba, Cyrille Adoula, Joseph Ileo, and several others. Meanwhile, in Elisabethville, during the first half of October, the Confédération des Associations Tribales du Katanga (Conakat) was organized under the presidency of Godefroid Munongo, who stepped down in favor of Moïse Tshombe after Conakat had been officially turned into a political party (July 11, 1959).[61]

Apart from its moderate leanings and its willingness to accept the affiliation of European groups (which was not officially done until the spring of 1959), Conakat was the type of organization that was likely to meet with the approval of the Belgian authorities. Membership was limited (at least initially) to ethnic associations representing Katanga tribes, and thus provided a multilateral structure in which differences between two or more Katanga ethnic groups could be mediated. All major ethnic associations in Katanga (except Atcar) rallied to the Conakat during the first few months of its existence. Even the Baluba-kat (Association des Baluba du Katanga), founded in 1957 by Jason Sendwe and later to become the nucleus of anti-Conakat opposition, joined in February 1959. The other member associations were the Groupement des Associations Mutuelles de l'Empire Lunda (Gassomel), led by Moïse Tshombe, the Fédération des Tribus du Haut Katanga (Fetrikat), the Association des Bena Marungu, the Alliance des Bahemba du Katanga (Allibakat), the Association des Basonge, and the Association des Minungu. The Union Katangaise, the leading European party, which had been founded in May 1958, became informally affiliated with Conakat in March 1959 and was officially admitted in June, but as the settlers became increasingly involved with the organization (and the financing) of Conakat, Sendwe and his followers, who were already concerned that the association's violent opposition to the Luba immigrants from Kasai might be extended against them, gradually drifted away from Conakat.

The issue of whether the Congo should be given a federal or a unitary structure rapidly became central in the mounting debate between Sendwe's groups and the Conakat. In May 1959, Conakat announced itself for an "autonomous and federated State [of Katanga] in which the reins of command will have to be in the hands of authentic Katangese," adding that "the *sine qua non* of the creation of a federal Congo lies in the equitable representation of each autonomous state in proportion to its economic importance."[62] Vice-Governor Schoeller, himself favorable to the idea of federalism, recognized however that

> if we opt for this system, we must also assure serious guarantees to the minorities. I am thinking, notably, of the very important Kasai minority in Katanga (about 38% of the population and 50% in some centers). This group fears, not without reason, a federalist regime under which "authentic Katangese" have expressed many times their intention of treating them as aliens. The opposition will come from these important minorities in the provinces of Katanga and Léopoldville rather than from the other provinces.[63]

This was exactly the feeling of the Kasai BaLuba community in Katanga, organized into an association known as Fedeka,[64] and, increasingly, that of the Balubakat as well. This was also, in a different context, the position of Atcar which, although made up of "authentic Katangese," had many good reasons to regard itself as a persecuted minority. On September 14, 1959, Balubakat came out in favor of a unitary form of government and invited residents of Katanga to relinquish all invidious distinctions between "strangers" and "authentic Katangese." This appeal was immediately endorsed by Atcar, which thereafter moved rapidly closer to Sendwe's group. After the latter formally broke all ties with Conakat on November 10, 1959, the way was paved for an electoral alliance between Balubakat, Fedeka, and Atcar, on the eve of the local government elections which the Belgian government insisted upon holding in December to gauge the representativeness of the various political formations before negotiating with them the terms of the Congo's decolonization. This alliance, which was not officially announced until *after* the elections, was to be known as the *Cartel Katangais*—or simply the Cartel.

The formation of the Cartel came just in time to rescue Atcar from

an increasingly untenable position. The efforts made by the administration to check the Lunda-Cokwe conflict had been hamstringing the
Cokwe movement by preventing it from evolving a coordinated organization. In Sandoa, Dilolo, and Kolwezi territories, Cokwe associations
had been recognized separately and allowed to have branches in each
chefferie (subject to the chief's intervention where issues of law and
order were concerned) but they were being closely watched by the local
administrateurs, who, at the end of 1958, could flatter themselves
(somewhat prematurely as it turned out) that

> we have witnessed in 1958 the gradual abatement of all the tensions
> that existed between Lunda and Tshokwe....The Tshokwe [of
> Dilolo and Kasaji] have been as level-headed and understanding
> ...as they had been aggressive and volatile before.[65]

The fact that "Ambroise Muhunga's visits have become less frequent,"
suggested the administrator for Dilolo (without mentioning that this
was largely due to the Belgian authorities themselves), was probably
one of the most important factors of this welcome development.

In fact, however, a new and more elusive form of infiltration was
already taking place throughout Lunda, and indeed throughout many
rural areas. This was the reverse flow of migration from the urban centers created by the economic recession which had begun to manifest
itself in 1957. Katanga was especially hard hit: in Elisabethville, the
number of African workers declined from 45,884 in 1956 to 37,950 in
1958 and the unemployment ratio grew from 4.8 to 13.6 percent
between 1956 and 1957. The number of industrial jobs, particularly,
dropped from 13,880 in 1956 to 8,273 in 1958.[66] Not every man who
was laid off left the city, of course. Many fell back on menial jobs, and
the number of domestic servants, for example, grew from 6,756 to
11,090 during the same two-year period. Even so, the African population of Elisabethville declined by 2,672 persons in 1958, for the first
time since the depression.[67] Jadotville and Kolwezi were similarly
affected.

Most of the workers who had migrated to the Copperbelt centers
from the Lunda heartland came from Sandoa or Dilolo territories,[68]
and this is where the majority of the reverse migrants now returned,
bringing with them not only their socio-economic frustrations but also
an impatience of traditional restraints which prompted many of them
to settle in villages not too distant from local sources of potential

employment. The chefferies of Sandoa and Dilolo territories increased their combined population by 17,289 units between 1956 and 1958 — over four times the increase of the previous two years — but virtually all of this growth was concentrated in the chefferies surrounding Sandoa, Dilolo, Kasaji, and Kisenge, to such an extent that Chief Lumanga, who had seen the population of his chefferie increase by 11.5 percent in 1958 alone, systematically refused to admit more newcomers.[69]

The return of these volatile elements dissolved the futile barriers with which the administration had sought to insulate the rural areas from "contamination" by city-bred ideas, and though neither the Lunda nor the Cokwe migrants had been notable for their activity in the incipient nationalist movement, it infused the rural communities with a new dose of militancy.[70] There were fresh incidents during the spring of 1959, notably at Kasaji, where Chief Tshisangama and two of his headmen tore down the fences that Cokwe villagers had built around their fields, on the grounds that this constituted an inadmissible sign of appropriation of a land over which they had no permanent rights.[71] Chief Samutoma, for his part, intensified his claims to be recognized as paramount chief of the Cokwe. His insistence on this issue was undoubtedly motivated by the increasing attentions being showered upon the Mwaant Yaav by the Belgian authorities since the formal launching of the decolonization process in January.

The Lunda paramount had at that time handed a highly publicized note to the colonial minister in which, speaking for the first time on behalf of the "Lunda Empire," he had expressed reservations about the introduction of universal suffrage in regions which, like Lunda, had had "for centuries a well hierarchized monarchic system parallel to that of certain civilized countries."[72] This position, which corresponded to the views openly expressed by groups of Belgian veterans and settlers and privately shared by many in the colonial civil service, had made the Mwaant Yaav a favorite of those European circles who opposed the new policy inaugurated by the Belgian government.[73] When the colonial minister, Maurice Van Hemelrijck, again visited Elisabethville in June, Ditende led the delegation of traditional chiefs who addressed him and, on that occasion, invited him to Sandoa to see "the real Congo" at first hand. The Mwaant Yaav joined the minister's official party on the flight to Sandoa where, on June 19, a spectacular gathering of chiefs was staged for Van Hemelrijck's benefit. Chief

Samutoma, who had been invited to attend, insisted upon resorting to the very same ceremonial (notably the use of a litter) as the Mwaant Yaav, who was persuaded only with difficulty not to create an incident in the presence of the minister.[74] Samutoma, on the other hand, affected to believe that this episode had entailed his formal recognition by the authorities as the Mwaant Yaav's equal and he renewed (unsuccessfully) his earlier demand to have his chiefly title officially changed to Mwatshisenge.[75] The next bout in this exercise in "one-upmanship" came on August 1, when Senator and former Prime Minister Joseph Pholien, a man of known conservative views, visited Sandoa as chairman of a senate fact-finding commission and was handed a memorandum by the Mwaant Yaav "on behalf of the 300,000 Lunda and Tshokwe of the Lunda Empire," whereupon the "Mwatshisenge" rose angrily and denied Ditende's right to speak for the Cokwe. The very next day, the two chiefs were on their way to Kolwezi to protest each other's behavior to the district commissioner, but each of them also alerted their supporters in Elisabethville, and on August 3 Moïse Tshombe (who had now become president of Conakat) and Ambroise Muhunga both appeared in Kolwezi to back up their respective sides in the dispute. The district commissioner refused to see the two urban leaders but otherwise espoused the Mwaant Yaav's side almost entirely. Samutoma (who had brought the other Cokwe chiefs along with him) was publicly reprimanded, suspended without pay for one month, and forbidden to leave Sandoa territory or to interfere in the affairs of any other chefferie under pain of dismissal.[76] The Mwaant Yaav's triumph was complete and he chose to interpret the episode as "a first step toward the restoration of his empire."[77] The discomfited Cokwe chiefs soon left Kolwezi, but not before Samutoma and Muhunga had been rudely jostled by a threatening Lunda mob which had collected outside the district office.[78] There were serious clashes between the Lunda and Cokwe communities in Kolwezi that night and during the next two days, and the district commissioner seized upon this pretext to bar Muhunga from entering the territories of Kolwezi, Dilolo, Sandoa, and Kapanga.[79]

From that moment on, any hope of reconciling Atcar and Conakat (assuming that either had entertained any such thought) was lost, and it was only a matter of time before Muhunga would join Sendwe's Balubakat in forming an anti-Conakat front.

But this was not the only card Muhunga could play to escape the administration's attempt to cut him off from his base of support in the southwest. Through his Luba connections in Elisabethville, he had come into contact with the local branch of Lumumba's MNC which was, at that time, trying to contain the effects of a disastrous split initiated in July 1959 by Joseph Ileo, Cyrille Adoula, and a few others, but later identified with Albert Kalonji (the future "Mulopwe" of South Kasai). The MNC's initial base in Katanga had been among the urban community of (predominantly Luba) immigrants from Kasai, but as Albert Kalonji emerged as leader of the MNC's dissident wing, his emotional appeals to the tribal solidarity of all Kasai BaLuba in their feud against the Bena Lulua had won many members of that community in Katanga over to his side and weakened the position of Lumumba's party.[80] In its search for a more diversified ethnic base in Katanga, the MNC, which had, like Atcar, endorsed the Balubakat manifesto of September 14, extended an invitation to Muhunga to attend its national congress in Stanleyville in October. The congress was held from October 23 to 26 but was followed on October 30 and 31 by serious riots leading to the arrest and imprisonment of Patrice Lumumba.[81] Muhunga himself was reportedly elected vice-president of the Katanga section of the MNC at Stanleyville but, significantly, made no mention of that fact in the account of his trip which he later addressed to the Tshisenge branch of Atcar.

[I met] strangers from Ghana, Guinea, Tanganyika, Uganda, Urundi, Scotland[?], Congo . . . . All said that the Whites must leave us alone. From the 22nd to the 30th of October, they wrote to the great chief of Belgium asking him to send a white man to Stan-[leyville] who can listen to all that we will discuss in our congress: we want to reach an agreement with you about ruling [i.e.: independence] and about choosing our men [i.e.: the December elections]. The Minister refuses. He says it is difficult to meet with you to discuss all that, all you need to do is to execute what we are telling you clearly, "a slave does not talk back to his superior." Thereupon the people said: in that case we will not choose our men [= we will boycott the elections]. On October 30 at 5 p.m. the Whites started shooting at us with their guns, first with tear gas then with bombs and submachine guns. Lord! how many people died on that day. Through His will I am not dead or wounded and I thank Him today by saying: "Blessings be to God, blessed be His Holy Name."[82]

But far from mentioning any ties between Atcar and the MNC, Muhunga went on:

> I heard that Mr. K[apenda] [= Tshombe] came to you to deceive you, saying that Atcar no longer exists, that it is dead, everybody has joined Conakat. I say to you and I write to you: do not try to enter their meeting. Atcar still exists and it is even known in Europe. Every Tshokwe who will accept a card of Kapenda Tshombe's Conakat, we shall call him "Lunda" and we do not want him in our association any more . . . . We no longer want this region to be governed by others without any Tshokwe.[83]

It seems clear that the Cokwe leader was not prepared to submerge his movement's identity into the MNC but envisaged instead a tactical alliance that would ensure nationwide exposure for Cokwe grievances—possibly along the lines of the arrangements that were subsequently concluded in Kasai between the MNC-Lumumba and the small non-Luba ethnic parties: UNC (Bena Lulua), MUB (Mouvement de l'Unité Basonge) and COAKA (BaBinji, BaKete, etc.). But whether or not they took it at face value, Muhunga's alleged adherence to the MNC was not designed to encourage the Belgian authorities in Katanga to a greater tolerance of his political activities.

During the month of December 1959, local government elections were organized throughout the Congo. Though relatively meaningless in themselves, these elections had acquired considerable psychological significance both in the eyes of the administration and in those of the more militant nationalist parties (notably the Abako-PSA-MNC Kalonji cartel in Léopoldville province) who decided to boycott them because they feared, not without grounds, that their activism would be swamped by the votes of the politically docile rural masses.[84] They had all the more reason to be suspicious since the suffrage in rural areas was to be universal but not direct. Elected members of the rural *Conseils de Territoire* would be chosen by representatives selected in turn from each native administration unit through various procedures. The proportion of elected versus ex officio members of the territorial councils was also to vary widely from one area to the next.

In connection with the elections, Muhunga again traveled to Kolwezi with other Cartel members to organize his supporters and was promptly arrested by the Belgian authorities for violating their earlier order to stay out of the district. The leading Lunda politician, Moïse

Tshombe, was of course under no such restriction. Even so, ten out of the eighteen elected members of the Sandoa territorial council and an even stronger majority of the Dilolo council turned out to be Cokwe — a direct consequence, the district commissioner observed, "of their numerical superiority." The results came as such a shock to some that Chiefs Tshisangama and Dumba openly talked of resigning and of withdrawing to the Musuumb.[85] Yet, no real campaigning had taken place and none of the Cokwe councillors had explicitly used the Atcar party label.[86] As a matter of fact, one small group called Atkat (Association des Tshokwe du Katanga), allegedly sponsored by Chief Samutoma, had shown signs of inclining toward Conakat.

This particular episode only added to Muhunga's fear that the show of Cokwe strength at the polls might not automatically benefit the cause of Atcar, and confirmed his suspicion that the administration had been quietly working on the two most politically-minded Cokwe chiefs, Tshisenge and Samutoma. Tshisenge, the more malleable but also the less prestigious of the two, had already been persuaded not to associate publicly with Atcar, but his people were among the most active supporters of the party and, following the elections, increasingly forced him into a more militant attitude (at least outwardly). On the other hand, Samutoma, though still bristling at any suggestion of Lunda supremacy, had been chastened by the Kolwezi incidents in August and could perhaps be dissociated from Atcar by a judicious stick-and-carrot approach. When King Baudouin had visited the Congo shortly after the elections, Samutoma had been authorized to travel to Elisabethville to attend the ceremonies. On that occasion he had been vigorously wooed by the Cartel in the hope of securing from him an endorsement of their positions, but Samutoma had prudently eluded their embrace and shirked any commitment,[87] although he had agreed to sign an address to the king which presented various Cokwe grievances (including Muhunga's banishment from Lunda). Samutoma also reportedly met with Governor Schoeller at the time of the king's visit. There is of course no record of what transpired during that meeting, but a few days later three hundred members of Atcar gathered at Kolwezi heard a spokesman for the chief tell them that the governor had "ordered him to forbid the Tshokwe from affiliating with any political parties susceptible of causing disturbances in Katanga . . . [such as] the MNC or Conakat," adding that he did not wish his people to be accused of causing clashes in the province "because of

the behavior of some Tshokwe who are eager to seek their personal
advantage by dabbling in politics."[88] Thereafter, the gap widened not
only between Muhunga and Samutoma, but also between that chief
and the Cokwe militants of Sandoa who accused him of betraying the
national cause, just as chiefs Tshisenge and Kandala were being
attacked by the activists of Dilolo territory for being too conciliatory.
But if Samutoma's credibility as a potential leader of the Cokwe nation
was rapidly being eroded, Muhunga's capacity to rally the Cokwe was
not expanding to fill the void, partly perhaps because of his own
limitations but also, no doubt, because of the continued restrictions on
his organizing activities. Thus, as the Round Table Conference
opened in Brussels and as the Congo moved into the final stage of its
precipitate decolonization, the Cokwe were suffering from a serious
leadership gap which they never successfully closed.

   In the escalating conflict between Cokwe and Lunda, the latter undoubtedly had some serious advantages. Some of them of course were
legacies of the colonial past, which previous chapters have attempted
to explore. Others, however, were of more recent vintage — such as
Moïse Tshombe's fast ascending career as a mediator of European and
African interests, or the administration's almost obsessive desire to
conjure up the earlier, simpler Congo of the dignified chiefs against a
Congo made up of Kasavubus and Lumumbas.

   Ever since organized political activity had made its appearance in
the Congo — or indeed since this appearance had first been appre-
hended at the end of the war — the administration had been
increasingly looking toward the chiefs for a counterweight to the growing influence of the articulate évolués. In such a perspective, the bureaucratic criteria by which chiefs had been judged in the past — particularly since the depression — were gradually superseded by political
criteria centering on the degree of control that a chief could be
expected to have over his people, and of course on the extent to which
such control could be regarded as a known parameter in projections of
future policies. This meant, paradoxically, that as the gap between
urban and rural values and aspirations widened, European strictures
against the supposed anachronism of chiefly rule actually decreased
(African criticism, of course, followed an inverse ratio). Similarly, the
administration's long-standing reservations about large-scale traditional entities was now mitigated by new considerations. The logic
which had led a Belgian official to argue as late as 1951 that the

Mwaant Yaav's chefferie was "too large" to be competently administered was now being reversed in the light of the hypothesis that the
Mwaant Yaav's stabilizing influence could be most effective if it was
permitted to apply to the largest possible area.

Ditende, whose excellent relations with the administration during
the early years of his reign have already been mentioned, acquired
even more stature among official and settler circles following his
response to the government declaration of January 1959. In that
message, the Mwaant Yaav had censured the "unforgivable aber-
ration" of "considering the opinions which emanate from the urban
centers as representing the general feeling of this province" in the
name of the "Lunda Empire, one of the most important demographic
groups in Katanga whose sphere of customary authority extends
beyond the boundaries of the Congo [into] Angola and [Northern]
Rhodesia."[89] This assertion prompted an immediate query from the
governor of Katanga to his native-affairs advisers regarding the consistency of Ditende's claims and the feasibility of allowing the Mwaant
Yaav to visit his alleged "subjects" outside of Katanga. In its reply, the
service of native affairs noted that the Lunda paramount had recently
traveled to Northern Rhodesia to attend his daughter's ritual inaugu-
ration as chief of Chavuma, had been consulted on a chiefly succession
case in Kwango as recently as 1957, and that his authority "un-
questionably" extended also into Angola.[90] Within a few days, a
projected itinerary had been drawn up with the help of the administrator for Kapanga and submitted for approval to the governor. It
involved a journey across the Angolan district of Lunda from Vila
Teixeira de Sousa to Tshikapa (where the Mwaant Yaav would meet
with the Kiamfu of the BaYaka and possibly accompany him to
Kasongo Lunda, on the Kwango) and back to the Musuumb by way of
Luisa.[91] Provincial authorities immediately initiated a correspondence
with the Portuguese consulate general and with their opposite
numbers in the two Congolese provinces of Léopoldville and Kasai.
Responses from the latter two quarters were clearly less than enthusiastic, however, the provincial commissioner for Kasai expressed serious
reservations, indicated that the Mwaant Yaav would not be permitted
to visit any villages, and added, for good measure, that his budget
would not contribute a penny toward the projected trip.[92] Reactions
from Léopoldville were even more negative. The Mwaant Yaav's
request had come at a time when Kongo radicalism was running high

and when, in the absence of their more sober-headed leaders, Kongo extremists were openly demanding the "partition of Belgian Africa."[93] The governor of Léopoldville Province was frankly hostile to the project and alerted the governor general to the danger that

> this trip might be given a distorted interpretation by some for the purpose of spreading rumors relative to a revival of the Lunda empire, which rumors would only add to and reinforce those which are already circulating about the reconstruction of the Bakongo Kingdom.[94]

Members of the governor general's staff were equally preoccupied by the Abako factor: the BaKongo might be impressed, their reply suggested, with a demonstration of the fact

> that they are not alone and that, if they mean to reconstitute the Kingdom of Kongo, the Mwata Yamvo will insist, for his part, on reconstituting the Lunda empire . . . [but], good tacticians that they are, they will only use this as an argument to back up their thesis that the unity of the Congo is artifical and takes no account of historical realities. At this point, they might well be backed by all the major chiefs in the country.

The letter went on to indicate that the Mwaant Yaav should be permitted to travel only as a private citizen and should be left to handle the arrangements for his trip without any official intervention — the unspoken assumption being that Portuguese authorities would probably refuse him a visa.[95] This was communicated to the administrator for Kapanga, who in turn informed the Lunda paramount but when, less than a week later, Ditende managed the remarkable feat of bringing Minister Van Hemelrijck back with him to Sandoa, he approached the governor general directly about his projected trip and allegedly received from him personal assurances of official intervention on his behalf. This launched a new round of correspondence on the part of the puzzled Katanga officials who by that time, however, had themselves changed their mind about the whole idea and were now less interested in the scope of the Mwaant Yaav's influence outside Katanga than in seeing him use it within the Lunda heartland to reduce mounting ethnic tensions and to ensure the success of the forthcoming local government elections.[96] As expected, the Portuguese authorities vetoed the Angolan portion of the proposed trip and all that remained was the prospect of a meeting with the

Kiamfu which, the governor eventually advised, "should be postponed until next year."[97]

Yet, the intriguing notion that the Mwaant Yaav's influence might still radiate, however faintly, over a substantial portion of central Africa continued to preoccupy some Belgian officials, not to mention the Lunda court itself. Thus, in January and February 1960, both Ditende and the governor of Katanga wrote separately to various authorities in the Congo, in Angola, and in Northern Rhodesia to inquire about the number of Africans under their respective jurisdictions who continued to recognize the Mwaant Yaav's paramountcy.[98] Other Belgian officials, however, warned against the dangers of fanning the Lunda imperial mystique and noted that the policy of Cokwe autonomy inaugurated in the 1920s had been the real base of peaceful coexistence between the two groups. "Any urge or attempt by the Mwata Yamvo to restore Lunda hegemony," wrote the district commissioner for Lualaba, would only serve to further alienate the Cokwe, adding that the behavior of a man such as Chief Samutoma was only a "function of the expansionist designs of the Lunda." The December elections, he observed, had shown that the Cokwe could not be ignored and that the only way for Conakat to rally them would be "to grant them the importance to which they are entitled." But so long as the Mwaant Yaav remained identified with that party, the chances of a reconciliation were slight.[99] Other officials continued to insist (somewhat wishfully) that the Lunda-Cokwe dispute had been blown out of proportion by "extremists" on both sides and pointed out the fact that Samutoma and Chief Mbako, the Mwaant Yaav's son, had remained on excellent personal terms during this period of tension.[100]

In fact, just as Samutoma's uncompromising attitude had been stimulated by a combination of traditionalist and modernizing elements, Ditende's new militancy was apparently nurtured by sons of the traditional elite, the most educated of whom often seem to have been the most belligerently "empire-minded." Most of these men had been trained by the Methodist missions and several of them, like Oscar Mbundj or Amédée Chisol, were employed as teachers or medical assistants in mission schools and hospitals. Another segment of the modernizing elite consisted of the sons of prominent African trading families, many of whom had started in business as associates or employees of the late Jospeh Kapenda; this was true of the Mawawa family, who were the sons of Kapenda's brother Sakambol and were thus first cousins of the Tshombe brothers; of Jacob Munana, a matrilineal cousin of Joseph Kapenda; or of the Kangaji family, whose business interests were centered in Kasaji and Kolwezi. Still others were administrative assistants of the Mwaant Yaav, such as Gabriel Kanyimbu Kambol, Ditende's principal secretary, or Ernest Koji, the treasury clerk for the chefferie.

There was of course a considerable overlap between these various groups. The Tshombe brothers' claims to fame rested not only on their father's business achievements but even more perhaps on their matrilineal descent from Mbumba and Mushidi. The career of a man like Oscar Mbundj illustrates the intricacy of these cross-affiliations: a grandson of the late *Nswaan Mulopw* Mulolo (who had been bypassed in favor of Ditende in 1951 because of his advanced age), he was, like the Mwaant Yaav himself, a descendant of Naweej Ditende, and had married in 1956 one of Ditende's own daughters, which made him not only the Mwaant Yaav's son-in-law (and remote cousin), but also the brother-in-law of Moïse Tshombe. A mission schoolteacher in the fifties and a onetime store clerk for Joseph Kapenda, Mbundj was successively appointed administrateur de territoire, then district commissioner with jurisdiction over Lunda after independence. Later still, after Tshombe's fall from power, he embarked on a successful business carer which he has pursued to this day.[101]

Virtually every one of these men had at one time or another lived in, or at least traveled to, the urban centers of Kolwezi, Jadotville, or Elisabethville. Since 1958, many of them had been looking for leadership to the Groupement des Associations Mutuelles de l'Empire Lunda (Gassomel) which had been organized in the major cities of Katanga by men like Moïse Tshombe, Gédéon Kambaj, Dominique Diur, etc. They tended to reproach the Mwaant Yaav for not having reacted with sufficient vigor to the first signs of the Cokwe national awakening, and apparently felt that Ditende, who had turned sixty in 1958, was too old and too closely associated with the colonial order to effectively champion the Lunda cause, even though they appreciated the tactical value of the prestige he enjoyed with the Belgian authorities.

Ditende's deep-seated suspicion of the Tshombe family, last evidenced in the *Nswaan Murund* incident, and in his diffident attitude toward Gassomel, was now fuelled by the appearance of Conakat which, in his eyes, was not so much a Katangese party, or even a

Lunda party, as it was the Tshombes' party. For the Mwaant Yaav to endorse Conakat, however, was probably the only way, at least in the short run, to avoid being outflanked by the younger Lunda activists who, in turn, realized that they could not afford to alienate the symbol of Lunda unity. The imperial mystique, whether viewed in terms of traditional prerogative (as it apparently was by Ditende), or in more instrumental terms of power, provided a common ground between the two camps. Men like Oscar Mbundj or Gabriel Kanyimbu Kambol, who had links with both sides and helped organize a Conakat party branch in Kapanga territory, served to bridge the gap between the court and the activists — or, as the administration tended to believe,[102] to manipulate the Lunda paramount in the direction of greater intransigence. Another possible catalyst in the emergence of the imperial vision may well have been the appearance in late 1957 of Daniel Biebuyck's study on the bases of Lunda political organization, which that noted anthropologist concluded with these words:

> Lunda political organization is a solidly built structure which it would be difficult to replace with a better one and whose replacement might in any case have unfavorable repercussions . . . . With some people having a centralized state organization like the Zande or the Kuba, we have been capable of utilizing existing institutions. Why has our attitude been different in the case of the Lunda whose institutions lent themselves admirably to the achievement of a harmonious whole? It may be too late to take advantage of the institutional complex offered by the Lunda, but there is still time to safeguard and to integrate the best of it.[103]

This factor was compounded by the presence at the Musuumb in 1959-1960 of a brilliant young anthropologist, Fernand Crine, who (much to the annoyance of the administration) supplied the Mwaant Yaav and his entourage with a good deal of the theoretical and scientific ammunition they needed to enhance the credibility of the imperial concept.

At any rate, the term "Lunda Empire," which had scarcely if ever been mentioned in public prior to January 1959,[104] rapidly became a household word among militant Lunda circles during the following months and clearly contributed to the polarization of Lunda-Cokwe antagonism by cementing the solidarity of Cokwe chiefs and commoners. As the tension mounted, the administration continued its somewhat desultory efforts to achieve a reconciliation. Belgian field officials meeting at Kasaji in January 1960 settled on the idea of holding a general convention of all chiefs of the Lunda heartland as a possible way of resolving ethnic differences. Instead, a semi-improvised meeting, attended by the Mwaant Yaav and eight other chiefs,[105] was held in mid-February in the Kisenge offices of the BCK-Manganese mining company, but resulted only in a further display of antagonism, not only between Lunda and Cokwe, but also between Samutoma (whose latest actions were now being regarded as traitorous by Atcar) and the other Cokwe chiefs.[106]

When the disgruntled Mwaant Yaav returned to the Musuumb on February 15, he found his entourage in an uproar over the latest of Muhunga's anti-Lunda salvoes. This was an interview granted by the Atcar leader to the Léopoldville magazine *Pourquoi Pas? Congo* on February 6, in which he declared:

> The Lundá—what are they? A handful of men. Ten thousand perhaps . . . . As a matter of fact, who are they, those Lunda? I will tell you: slaves! Our former slaves! . . . After independence we will take care of them. We shall see who is stronger.[107]

This insulting outburst was immediately challenged by the Lunda association of Elisabethville in the form of a complaint to the public prosecutor's office,[108] but the outrage felt at the Lunda court ran far deeper. Yet, the peremptory letter that the Mwaant Yaav fired off to the governor general on February 25 made no reference to the Muhunga interview but only to "the growing tension between Aruund and Tshokwe in the region of Sandoa" and was almost entirely devoted to an impassioned denunciation of Samutoma's pretensions and of the Belgian administration's disregard for his grievances. Half of the letter was taken up by a scholarly disquisition (presumably concocted with Crine's assistance) on the historical, political, and customary law aspects of the Lunda-Cokwe problem as viewed from the Musuumb, to which was added a somewhat less scholarly but revealing two-page appendix, listing no fewer than thirty-two tribes and sixty major chiefs claimed as traditional subordinates of the Mwaant Yaav. The list included every group that had ever had a Lunda chief or migrated from the Nkalaanyi over the previous four hundred years from the Imbangala to the BaBemba, and grandly embraced such important chiefs as the Kiamfu of the BaYaka, the Citimukulu of the BaBemba,

and of course the Mwatshisenge of the Cokwe (not the "usurper" Samutoma, but the one from Angola).[109]

The administration prudently declined to comment on the historical
or legal merits of the Mwaant Yaav's brief, arguing (as colonial rulers
are wont to do in a terminal colonial period) that the matter was best
left for the government of an independent Congo to decide, but they
were concerned by the letter's somewhat cryptic demand that a "firm
solution" be found "irrevocably before March 15." What did the
Mwaant Yaav have in mind, the administration wanted to know, and
more important, what did the Lunda plan to do if their ultimatum was
not met? The assistant district commissioner was immediately dispatched to the Musuumb and from a series of meetings with the
Mwaant Yaav (alone and with his council) gathered that the Lunda
paramount would settle for an act of ritual obeisance to be performed
at the Musuumb by Chief Samutoma.[110] As to the deadline of March
15, the Mwaant Yaav was evasive, but at that same time, in Dilolo
territory, the rumor was already current that on that day, or even
sooner, "there would be war" between the Lunda and the Cokwe.[111] In
fact, the few incidents that did break out were overshadowed by the
widespread riots of March 12 and 13 which coincided with Lumumba's
visit to Elisabethville. For two days, in Elisabethville, Jadotville, and
Kolwezi, there was indeed war, not only between the Lunda and the
Cokwe but between followers of the Cartel and of Conakat, between
immigrants and "authentic Katangese," between BaLuba and Lulua,
and even between Katanga and Kasai BaLuba.[112] With little more
than a hundred days left before independence, the colonial system was
giving way to a kaleidoscope of passions and interests from which
would emerge not a Lunda empire but a State of Katanga, whose life
and death would condition the fortunes of the Mwaant Yaav's
kingdom.

8

Decolonization, Secession, Reunification: The Lunda and Cokwe Through the Congo Crisis

The last months of Belgian rule were characterized by a widening gulf between illusion and reality, and between theory and practice, on the side of the colonial authorities as well as on the part of the Africans. From the opening of the Brussels Round Table Conference to the formation of the first national government of the Congo, Belgian illusions regarding the stabilizing influence of rural moderates and African illusions of nationalist unity were gradually shattered. At the same time, both sides were waging a losing battle to reconcile their mutual desire to maintain the facade of an orderly transfer of power with their increasingly manifest lack of any real control over political developments throughout the land. Wildcat strikes in Katanga, ethnic conflicts in Kasai, civil disobedience in the Lower Congo were only a few of the signs of disintegration apparent through the spring of 1960.

Despite the rise of ethnicity, few chiefs, whether in Lunda or elsewhere, were in a position to exercise concerted political influence. As he returned from the Round Table Conference which he had attended as one of the ten members of the chiefs' delegation, the Mwaant Yaav's son, Chief Paul Mbako, could perceive, better than most of his peers, the dilemma of the traditional rulers. Although they had originally been granted ten of the forty-four African seats at the conference, the chiefs had been talked into accepting a voting procedure whereby each of the eleven delegations (some numbering as few as one or two members) would speak with a single voice, leaving them outnumbered in most cases where a split decision had been reached. More to the point, the chiefs had found themselves siding over every issue with Conakat and the PNP (Parti National du Progrès), a hetero-geneous assemblage of moderates with a well established image of sub-servience to the administration.[1] With no program of their own, the chiefs thus found themselves not only espousing a pro-administration stance, but also being led by, rather than leading, the other "moder-ates." On the other hand, those chiefs who, at the Round Table or back in the Congo, were willing to side with an established party often discovered that their prestige value was being utilized by urban *évolués* over whom they had little control.

On the basis of his experience in Brussels, Chief Mbako apparently reached the conclusion that a posture of strict nonpartisanship (at least for the time being) was the only way for chiefs to live down their past record as auxiliaries of the colonial system and to preserve their image as community leaders in a time when political passions were becoming strongly polarized. Whether he fully realized at the time the potential risks of a close identification with Moïse Tshombe (who had repeatedly found himself at odds with other Congolese delegates in Brussels) is doubtful, but the fact is that, upon his return, he threw his consider-able influence behind the idea of keeping party politics as much as possible out of Sandoa territory—an endeavor in which he was sup-ported by the local administrateur and (according to one informant) by the Methodist mission. Local chiefs, most of whom had little to gain from an exacerbation of Lunda-Cokwe tension, were easily persuaded. Samutoma, who had now been ostracized by Atcar for his divisive tactics and who was steering his own son, Célestin Sapindji, into the political arena,[2] was also won over to Paul Mbako's position, especially after Moïse Tshombe's brothers, David Yav and Thomas Kabwita, agreed to the terms of the compromise. As a result, party labels—if not ethnicity itself—were dispensed with in the provincial and national elections. Four candidates (two Lunda and two Cokwe) ran for the two seats allocated to Sandoa territory in the Katanga legislature: these were Paul Mbako, Thomas Kabwita, Célestin Sapindji, and Antoine Itunga, all officially running as "independents." Mbako and Kabwita were elected to the Katanga Provincial Assembly while the "nonpar-tisan" vote from Sandoa ensured the election of Chief Muteba to one of the three seats from the District of Lualaba in the national House of Representatives (under the label of "Autorités Coutumières du Katanga"). The fact that all three subsequently rallied to Conakat and to the secession shows the limit of their nonpartisanship, but the whole

operation had the merit of defusing tensions during a crucial period —
and, from a Lunda viewpoint, of undercutting Atcar's expansion in an
area where the Cokwe represented a numerical majority. Moreover, by
avoiding explicit affiliation with any one party, Paul Mbako's compromise formula left open the possibility of a regional, rather than
ethnic, identification which, if extended to the two other territories,
might have preserved the Mwaant Yaav's prestige throughout the
Lunda heartland far more effectively than the highly controversial
"empire" concept.

Whether the Mwaant Yaav failed to perceive the long-term implications of his son's initiative or whether he feared that a clear-cut polarization might expose the alienation of some of his subjects, there was
no question of giving up party labels in Kapanga territory and no
organized political activity outside Conakat. This was all the more
striking since the area was ethnically the most homogeneous in the
Lunda heartland. Yet, traditional as well as more recent forms of
factionalism were reflected in the fact that no fewer than four separate
candidates, all purportedly carrying the banner of Conakat, were
entered to compete for the two seats allocated to Kapanga in the provincial assembly.[3] Ernest Koji, who was known to have Ditende's
implicit backing, and Oscar Mbundj, representing the orthodox or
"Tshombist" version of Conakat, were both elected, with 33.45 and
36.58 percent of the vote, respectively (see Table 18). It was only in
Dilolo territory that Atcar found itself in a position to capitalize on the
depth of ethnic cleavages by carrying two out of four provincial seats
with 59.4 percent of the vote, but Conakat held its own among the
Ndembu and Lwena. The Conakat also won two out of three seats
from neighboring Kolwezi territory with 55.8 percent of the vote,
versus 36.7 percent to the Cartel.[4]

In the national legislative elections, three of Katanga's sixteen seats
in the 137-member House of Representatives were filled from the District of Lualaba, serving as a multimember constituency. The Atcar
slate carried 41.2 percent of the vote and Conakat only 23.2 percent,
while 35.6 percent (most of it from Sandoa territory) went to the
supposedly nonpartisan ticket labeled "Autorités Coutumières du
Katanga" but the three slates nevertheless elected one member apiece,
the Atcar seat going to Ambroise Muhunga.[5] Thanks to Paul Mbako's
neutralization of Sandoa territory and to the administration's repeated
obstruction of its activities,[6] Atcar had finally polled no more than

Table 18. Katanga Provincial Assembly election, 1960, by district and (Lualaba only) by territoire.

	Conakat	MNC/K	Union Congolaise	Cartel Balubakat	Cartel Katangais	Atcar	Cartel Baluba-MNC/L	MNC/L	Local interests	Individuals	Total
District of Lualaba											
T. Kapanga	2										
T. Sandoa									2		
T. Dilolo	2					2					
T. Kolwezi	2			1					1		
T. Lubudi											
Total	6			1		2			3		12
District of Tanganyika	3		1	6			1	1	2	2	16
District of Haut-Lomami	3			9					1		13
District of Haut-Katanga	9									1	10
City of											
Elisabethville	3	1			2						6
Jadotville	1			2							3
Total Katanga	25	1	1	18	2	2	1	1	6	3	60

Cartel: 23

Source: Jules Gérard-Libois, *Katanga Secession,* p. 66.

two-thirds of the Cokwe vote and Muhunga had some reason to be embittered as well as disappointed by the modest results achieved by his party. This feeling was shared by other Cartel leaders when they discovered that although they had polled more votes than Conakat, over the whole of Katanga, the latter party would have twenty-five seats against their twenty-three in a provincial legislature consisting of sixty members. Their frustration grew during the following days, as Conakat managed to rally all the supposedly "independent" members (whose sympathies for Conakat had in fact often been transparent) and even threatened to disrupt the Cartel itself by concluding an agreement with Albert Kalonji, who might swing one or two Kasai Luba assemblymen away from the Cartel.[7]

By the time the Katanga provincial assembly opened its first session on June 1, Conakat appeared to control thirty-eight out of sixty votes, and the Cartel was reduced to the obstructionary tactics of boycotting the session in order to prevent the legislature from achieving the two-thirds quorum required for it to elect a provincial government. With the legal assist of an amendment to the Loi Fondamentale hastily voted by the Belgian parliament, this obstacle was eventually circumvented and a Conakat-dominated cabinet of ten members, presided over by Moïse Tshombe, was elected on June 16. The previous day, upon learning about the vote of the amendment in Brussels, Sendwe and Muhunga had finally committed the Cartel to an alliance with Lumumba — who had been assiduously courting them for over two weeks — for the purpose of building a national coalition government. Actually, when the Lumumba cabinet was elected by the national parliament on June 23-24, it included two Conakat members (minister of economic affairs and undersecretary of defense) and only one from the Cartel (minister of justice), but the Cartel had also secured for Jason Sendwe the potentially crucial post of state commissioner for Katanga, which it hoped would serve to compensate it for its exclusion from the provincial government.[8]

From Elisabethville, Tshombe reacted with bitterness to Sendwe's proposed appointment and declared that his party no longer considered itself bound to the coalition. The following day, a first attempt at secession was foiled by the Belgian authorities, but less than three weeks later (July 11), after Belgian troops, unrequested by the central government had been landed at Elisabethville, Tshombe declared the "total independence" of Katanga.

The secession was far from being entirely improvised, of course, and it benefited powerfully during its initial period from crucial military and technical assistance extended by Belgium, as well as from the emotional backlash created among expatriate circles by the uncontrolled violence loosed against Europeans in many parts of the Congo. The swift consolidation of the Katanga regime placed the Cartel opposition in an almost untenable position. While most major Conakat personalities had elected to serve in the Katanga provincial assembly, in conformity with their regional and federalist inclinations,[9] the leaders of the opposition (Sendwe, Muhunga, Rémy Mwamba, Isaac Kalonji) had for the most part sought national office and were now stranded in Léopoldville, out of touch with local party cadres and with their own disoriented followers. A small but significant number of local opposition personalities were enticed or intimidated into rallying to the secession; others were rounded up and placed in detention. In the solidly Balubakat regions of north-central Katanga, however, grass-roots opposition to the Elisabethville regime soon developed into a home-grown yet murderous form of guerrilla warfare which the secessionist government repeatedly but vainly tried to eradicate.

The fact that this area was contiguous to regions which continued to recognize the Léopoldville government (or, after November 1960, its Lumumbist rival in Stanleyville) made it possible for Balubakat leaders to use it as a base, albeit a precarious one, from which to challenge the Katanga regime on its own ground. A "Province of Lualaba"[10] was declared in October 1960 and later installed at Manono with the support of Lumumbist troops from Stanleyville. After the protracted negotiations between Léopoldville and Elisabethville had ended in deadlock and the tide began to turn against Tshombe, the Katanga regime had to resign itself to the virtual loss of the north where a "Province of North Katanga"—this time under the jurisdiction of the Adoula government in Léopoldville—was officially recognized in July 1962.

The Cokwe component of the Cartel was not so fortunate. Atcar's zones of popular support lay well beyond the reach of Congolese troops, insulated from other opposition centers by regions loyal to the Tshombe government and thus easily controlled by Katanga forces. Once more, Muhunga's ironic fate was to be cut off from his followers and condemned to survive politically as a protégé of his Luba allies, having outlived his usefulness before the end of the secession. He was given a rather nebulous position which he apparently filled only on paper in the Balubakat "Province of Lualaba,"[11] then was named undersecretary of state for mines in the Adoula cabinet in August 1961, only to lose his post in an obscure reshuffling after less than a year in office. By the time the secession came to an end, Muhunga's political stature had shrunk considerably and he never thereafter regained his former position of influence.[12]

Cokwe activists were persecuted, although Muhunga's claim made from exile that "thousands of men are being killed in Dilolo territory"[13] was an obvious exaggeration. In the urban centers of Elisabethville, Jadotville, or Kolwezi, Cokwe residents were often involved on the side of the BaLuba and of the Kasaians in clashes with the police or with segments of the population favorable to the Tshombe government. As the regime became more concerned with the danger of urban unrest (which, they feared, might give the United Nations an excuse to intervene) and more ruthless in its handling of dissidents, many of the Cokwe in Elisabethville sought asylum with the United Nations and were herded together in the notorious "Baluba Camp" — where again, as in the Cartel, they were eclipsed by their more powerful associates.[14] Many of these refugees turned to occasional banditry as a means of escaping the appalling living conditions of this sprawling ghetto, while others simply squatted in the outlying residential sections of Elisabethville, which had been deserted by their European occupants. By the summer of 1962, however, most of the refugees had been repatriated or had gone back to living in the townships. In the rural areas of the Lunda heartland, control of the Cokwe population was left largely in the hands of local authorities, especially the chiefs.

The role of the chiefs during the secession is one of the least documented facets of the Katanga story. From the outset, the Tshombe regime sought the endorsement of the chiefs for its decision to declare independence from the Congo. In accordance with the provisions of the Loi Fondamentale, the fourteen-member representation to which each province was entitled in the national senate was to be elected by the provincial assembly and had to include a minimum of three traditional chiefs.[15] Katanga, for its part, had elected four of them: the Mwaant Yaav, Chief Tshisenge (of Dilolo), Chief Katanga Kianana of the BaLemba, and Paramount Chief Kabongo, head of the largest Luba chefferie in Katanga. Their sponsorship was explicitly invoked during the extraordinary session of the Katanga assembly which ratified the declaration of independence on July 17.[16] Meeting in Elisabethville the previous day, these four chiefs and four of their peers[17] had issued a "solemn proclamation" approving "unreservedly" the independence of Katanga. Concomitantly, Tshombe had elevated ten major chiefs to the rank of "Ministers of State."[18]

In return for their loyalty to the Katanga cause (which had of course never been in doubt on the part of some of them), the chiefs were given a choice place in the new institutional structure of the Katanga state. Instead of a "senate" that might accommodate the now superfluous representatives of Katanga in the Léopoldville parliament, along with the new "ministers of state," the Katanga constitution promulgated on August 5, 1960, introduced a Grand Council consisting of twenty traditional chiefs selected by their peers and whose advice over certain issues could be contradicted only by a two-thirds majority vote of the National Assembly.[19]

On that occasion, the chiefs were also promised a restoration of their authority and a share of executive power at the local level. What this meant in practice however, was never fully spelled out. During January 1961, the heads of several branches in the Department of the Interior — along with their expatriate advisers — met repeatedly to study this question and to draft a statute that would specify the functions and prerogatives of traditional chiefs within the overall reorganization of the state. From the start, the discussion took an ambiguous turn as a result of the insistence by Munongo's Belgian *Chef de Cabinet* that while the intended reform should restore to the chiefs their "political powers," these did not really need to be defined. With a fine disregard for the record of Belgian native policy, the *Chef de Cabinet* declared:

> The purpose of the administrative reform is to restore to the chiefs, at the chefferie level, their administrative powers and above all their *political* powers. Before the Europeans arrived and until the Decree of 10 May 1957, administrative and political powers belonged to the chief surrounded by his notables. The power of the chief surrounded by his council is absolute, autocratic. Traditional authorities existed. They still exist. They are accepted by the population. Their responsibilities and functions are known and accepted by the population. Thus we do not have to preoccupy ourselves with them. As regards organization at the chefferie level, we are proceeding from a given that already exists.[20]

Repeated queries as to just what was meant by these "political powers" were met with an embarrassed silence or with the awkward reply that "such a definition is impossible."[21] One thing at least was certain, the two senior Belgian advisers insisted: the democratic system at the chefferie level was a failure and should be excluded. This would mean the disbanding of the partly elective *Conseils de Chefferies* which the colonial administration had timidly tried to introduce during the last two years of Belgian rule. This was questioned by several African members. In fact, throughout these discussions, there was a strange cleavage between the European advisers who, with obvious nostalgia, rhapsodized about the precolonial age when chiefs had "the power of life and death" over their subjects, and the African officials who, as urban *évolués*, were concerned about possible abuses of chiefly power and about the status of educated commoners and "strangers" under such a system.

This was not the only instance in which educated Africans, who now filled most of the intermediate ranks of the Katanga state administration, expressed reservations about the idea that chiefs should be granted any exorbitant prerogatives. In the field, newly appointed African administrators who had replaced some of their Belgian predecessors[22] often complained that local chiefs not only refused to accept their directives but even demanded of them the same marks of servile respect as from their own subjects. This was particularly infuriating to the Africans who remembered only too well how subservient the chiefs had been to the colonial administration.

Actually, the Conakat leaders themselves, who were primarily educated urbanites even though many of them had ties with the traditional establishment, did not have the slightest intention of turning Katanga over to the chiefs. Godefroid Munongo, who had graduated from the Kisantu School of Administration and had been one of the first Congolese to serve as an *Agent Territorial,* was imbued with an almost Hegelian sense of the State. As for Tshombe, the ambivalent nature of his relationship with the reigning Mwaant Yaav probably tempered whatever inclination he may have had to build up the power of the Lunda monarchy.

Still, the chiefs were too essential to the internal order and security of the beleaguered state to be pacified with titles and trinkets at a time when the regime was being challenged on its northern flank by the consolidation of the Luba rebellion. Once already, the Grand Council had sided with the National Assembly in an attempt to assert parliamentary control over the presidency. This move was renewed on April 5, 1961, but the following day Tshombe finally decreed that chiefs, with their traditional councils, would be the sole authorities within their respective chefferies, to the exclusion of any democratically elected elements. This did not prevent Chief Mwenda Munongo (Godefroid Munongo's brother) from declaring on April 7 that since the proclamation of Katanga's independence, "nothing had changed in the situation of the traditional chiefs" and from demanding salary increases and other benefits.[23]

This latter issue remained a bone of contention between the chiefs and the government through the end of the secession. Two months before the capitulation of the Katanga regime, the Mwaant Yaav was still complaining to Munongo that "it is deplorable to note that Paramount Chief Mwanta-Yanvo is still being paid to this day as the head of a chefferie and not as Emperor of the Lunda Empire."[24]

In fact, this was only one aspect of the uncomfortable relationship that prevailed in Lunda between the victorious Tshombist faction and the Mwaant Yaav during the last three years of Ditende's reign. Viewed from the Musuumb, Conakat's takeover of Katanga implied not so much the promise of a restoration for the "Lunda Empire" as the achievement of tremendous power and prestige by the Tshombe family, and thus a distinct challenge to the Paramount's position. Such blandishments as Ditende's appointment as minister of state and member of the Grand Council, or the promises of a restoration of chiefly authority offered during the first weeks of the secession, did not entirely reassure the Mwaant Yaav as to the nature of the new regime's intentions.

In the capital itself, the "young Turks" led by Daniel Tshombe, Oscar Mbundj, and others, were talking openly of the need for a changing of the old guard. One of the first actions of the new government had been to appoint Mbundj to the post of administrateur de territoire for Kapanga, from which position he could supervise the Mwaant Yaav. Ditende's personal "adviser," a former field official named Gomez, identified three sources of opposition to the government's ostensible policy of reinforcing the powers of the chiefs:

(1) The Methodist Mission which wants to impose its views on the Mwata Yamfu, as in the past. It does not allow its decisions to be discussed.

(2) The current Administrateur de Territoire . . . who wishes to rule

the territory in a dictatorial way and will not agree to relinquish the slightest initiative or even the slightest freedom of action to the Mwata Yamfu.

(3) A veiled and latent hostility on the part of M. Daniel Tshombe, who claims that the Mwata Yamfu (a) is not liked by a portion of his people; (b) has been in office too long and should be replaced.[25]

What apparently caused Gomez to suspect the Methodists (a suspicion which, ironically, was also shared by Mbundj, as his correspondence shows, though for different reasons) was their attempt to arrange with the State Department a trip to the United States for the Mwaant Yaav. In Ditende's absence, Gomez told Mbundj, it would be only too easy for the Katanga government to dethrone him and to replace him with a man of their choice[26] — presumably (though neither man said so explicitly) a member of the Tshombe clan. It seems of course unlikely that the Methodist Mission would have been a party to such devious plans (if only because Gaston Mushidi, the pretender backed by the Tshombes, was a Catholic), but the voicing of these suspicions is indicative of what must have preyed on Ditende's mind during the early months of the secession.

In fact, aside from his personal feelings toward Ditende, Moïse Tshombe was probably too cautious a statesman to run the risk of causing a rift among his Lunda followers simply to satisfy an old family feud — so long as the Mwaant Yaav was willing to place the weight of his prestige squarely behind the regime. It was only two years later, in October 1962, as Ditende's health declined and the secession itself was clearly on the wane, that Tshombe's uncle, Gaston Mushidi, was finally placed on the road to the throne by being made Chief Lumanga, while the incumbent Lumanga, Mutshaila Ditende, was elevated to the position of *Nswaan Mulopw,* which he still held in 1973.

Just as Ditende's usefulness to Tshombe would allow him to die on the throne, the idea of a "Lunda Empire" was indirectly kept alive, at least to some extent, as an arm of Katanga's foreign policy. Thus, the "historical" ties between the Mwaant Yaav and the BaYaka of Kwango (which had been the object of much correspondence in 1959) provided a useful background to the numerous contacts which the governments of Elisabethville and Léopoldville developed from the moment that Antoine Gizenga organized a Lumumbist counter-government at

Stanleyville in November 1960. It is a matter of record that the most active mediator on the Léopoldville side was Albert Delvaux, leader of the Luka party, the political arm of the BaYaka tribe.[27] Earlier still, when Katanga (loudly supported by most of the Belgian press) had announced its intention of acting as the rallying point of an "anti-Communist confederacy," the thought of utilizing the Mwaant Yaav's influence to secure Kwango's adherence to this scheme—improbable though it was—had apparently entered the mind of some. The Mwaant Yaav's "adviser," for one, had a plan:

> If the Republic of Katanga cannot act directly over these populations, whether Basalampasu or politically inclined toward Luka and PSA—all of them of Lunda origin [sic]—could it not be done through the Mwata Yamfu's intercession? . . . The unquestionable and unquestioned *spiritual* leader of all these people is and remains the Mwata Yamfu. . . . It would therefore be useful for the Republic of Katanga to make available to the Mwata Yamfu a credit . . . for the purpose of making political gifts to the major chiefs of these populations. . . . In return [Katanga] would be assured of finding allies if it ever needed them.[28]

Less far-fetched and strategically more useful to Katanga were the contacts that the Mwaant Yaav maintained with chiefs in Northern Rhodesia during two trips made in the dry season of 1961 and 1962, although it was clearly the policy of the white government of the Central African Federation rather than any sense of Lunda solidarity that made it possible for Tshombe to use Northern Rhodesian soil in his attempts to outwit the U.N. forces.[29] This was even more apparent in the case of Angola, which Tshombe used as a staging area for several military operations involving African troops as well as mercenaries, not only during the secession but also during his two subsequent periods of exile (the last attempt to subvert the Congo from Angolan soil being made in November 1967, when Tshombe himself was already in the custody of the Algerian government). What contacts were established in Angola took place with the Portuguese authorities, usually through one of the Tshombe brothers, and these obviously involved other matters than the Mwaant Yaav's claims of paramountcy over the Portuguese-held segment of Lunda.[30]

Perhaps the most authentic example of a continuing imperial preoccupation on the part of the Lunda court had to do with the northern boundary of the heartland. There, it will be remembered, several

Lunda rulers of past centuries had fought and died trying to extend their authority over the Asalampasu, BaKete, and Bena Kanioka. The vagaries of administrative policy had drawn a somewhat arbitrary provincial boundary between the Lunda heartland and some of its northern marches where local land chiefs had sometimes continued to pay tribute to the Musuumb over a number of years. Between 1921 and 1926, as a matter of fact, a minor conflict had developed between the provincial administrations of Kasai and Katanga about jurisdiction over portions of this border area. Although that particular dispute had eventually been settled against Katanga (i.e., against the then Mwaant Yaav Kaumba), contacts had continued with these "lost" communities and in the protocol attached to his letter of 25 February 1960 to the governor general,[31] Ditende had claimed traditional jurisdiction over most of this area. In the autumn of 1960 several Salampasu village chiefs traveled to the Musuumb to bring tribute, and there (according to Gomez) "asked to be integrated into the Republic of Katanga."[32] Later during the secession, the Mwaant Yaav was also reportedly approached by Kanioka chiefs from the Mwene-Ditu territory for the purpose of seeking inclusion into Katanga.[33] This unusual démarche was in fact part of a developing conflict over the Mwene-Ditu area between the two provinces of South Kasai ("King" Kalonji's former "Mining State") and Unité Kasaienne, both created on August 14, 1962. Some Kanioka groups who were unhappy with this alternative were now apparently looking to the south for a third choice, but it was not until after Katanga had been reintegrated with the rest of the Congo and partitioned into three provinces (North Katanga, Katanga Oriental, and Lualaba) that the Mwene-Ditu issue really came into its own, threatening at that time to upset the balance of forces in the southwest. We will return to that problem later.

On the whole, then, the balance of the secession years as far as Lunda imperial ambitions were concerned was almost entirely blank. It was only to the extent that Cokwe self-assertion had been curbed in the heartland itself that one might speak of a Lunda restoration. As the secession came to an end, however, it could be questioned whether even that limited achievement would be maintained. Ever since the central government had accepted the principle that the Congo's six provinces could be fragmented,[34] Ambroise Muhunga, from his Léopoldville exile, had been concocting plans for a "Cokwe province" to be carved out of southern Katanga, and there were some among the members of the central government who felt that a partition of Katanga would act as a barrier against further attempts at secession.

The fact that the political and administrative cadres of the secessionist regime were allowed to remain in office after Katanga had been forcibly reunited with the rest of the country, however, significantly altered the conditions under which partition was eventually achieved. From the beginning, Tshombe and his associates insisted that there could be only one Katanga and demanded the reunification of the northern and southern halves of the province. Although he was not followed by his Balubakat colleagues in North Katanga, Jason Sendwe himself favored the idea of a reunited Katanga. Within the former secessionist state, meanwhile, the frustrations of those opposition groups who had looked forward to a day of reckoning, which now seemed to be eluding them, led to several violent incidents (the most serious of them occurring at Jadotville and Kolwezi in early April 1963) but these were sharply repressed by the same police that had maintained order during the secession.[35]

Yet, Moïse Tshombe's personal position in Katanga was gradually becoming untenable. Since he was the best known symbol of the old regime, his political visibility was such that his continued presence in Katanga compounded the ambiguity of the situation and cast doubts on the central government's resolve to liquidate the sequels of the secession. Tshombe himself had sensed this from the moment of Katanga's capitulation, but he could not afford to withdraw from the scene without having made some arrangements that would safeguard his political future. The presence of Katangese gendarmes in Angola was one such form of insurance policy. Another was the political survival of some of his trusted associates. In early February 1963 Tshombe took a short trip to Angola, followed by a much longer one to Europe, from which he returned in the company of Dominique Diur, a former vice-president of Gassomel, who had been representing Katanga in Paris. But pressure against Tshombe was mounting. On May 24 his personal residence was searched and his guards disarmed. Concomitantly, the bill partitioning southern Katanga into two provinces, which Muhunga and others had been actively pushing in Léopoldville, was passed by both houses of Parliament.[36] At that point, the news came from the Musuumb that the Mwaant Yaav was dying.

Ditende's death on May 31, 1963, came at a delicate yet opportune time for the Tshombes, and the problem of succession was quickly dispatched. Although the council went through the motions of comparing the claims of the different royal lines,[37] there seems to have been no question that the throne would now finally pass to a member of Tshombe's clan. As dutifully recorded by the loyal Oscar Mbundj,

> the council of elders, presided over by the Kanampumba Yav, began with a determination of the different families of former Mwanta Yamvos. It was recognized that one of these families had reigned only once. Effectively, Paramount Chief Mwanta Yamvo Mushidi, who was succeeded by Muteba, was the first [of that line] to reign. To satisfy every family and to complete the dynasty, the council . . . gave its preference to Mushidi [i.e., Gaston], of the above-mentioned family. . . . It will be remembered that after the death of Mwanta Yamvo Kaumba, Mushidi had been in competition with the late Ditende for the succession as head of the empire. Accordingly, his current selection caused considerable satisfaction among the people, as evidenced by the ensuing comments, which were all favorable. It might also be mentioned, in passing, that in the event of a refusal by Mushidi to succeed Paramount Chief Ditende, the council of elders would have pronounced in favor of another member of the same line, just in order to complete the dynasty.[38]

Thus, after fifty-six years, the royal *lukano* had returned to the line that had been swept from the throne with the help of King Leopold's agents—the line of Mushidi and Kawele, the heroes of Lunda resistance to alien rule—with the ironical twist that this restoration was largely due to a man who himself owed much of his power to a later breed of colonialists. In addition to Gaston Mushidi's accession to the dignity of Mwaant Yaav, arrangements were also made for Moïse Tshombe's brother, Thomas Kabwita, to succeed his uncle as Chief Lumanga. When Tshombe returned from the Musuumb, it was only to leave Katanga, this time for a year—but with the firm intention of coming back when the time was ripe. As he flew from Salisbury to Paris, then to Spain, he could doubtless reflect that his own political interests, not to mention those of his family, would be well looked after in his absence.

The organization of the two provinces carved out of the Conakat stronghold of southern Katanga (Lualaba and the southeastern rump of Katanga Oriental) seemed to confirm—at least initially—the resilience of Tshombe's influence. In Katanga Oriental, to no one's surprise, a Conakat government headed by Edouard Bulundwe[39] and including leading secessionist personalities such as Munongo and Kibwe was elected on August 13. But in Lualaba, which Muhunga had so confidently expected might become a "Cokwe province," Conakat also elected nine of the eleven members of the cabinet on September 29, including Thomas Kabwita and Dominique Diur, the man Tshombe had brought back from Paris in March and who now became provincial president. Only two cabinet posts went to Atcar. Even in North Katanga, Sendwe's wing of Balubakat, which had moved closer to Tshombe's position on several issues during the early months of 1963, managed to wrest temporary control of the government from the veterans of the antisecessionist guerrilla, led by Prosper Mwanba-Ilunga.[40]

From an early date, however, there were signs that neither Diur nor for that matter Bulundwe intended to be a mere instrument in the hands of the secessionist old guard. Diur, whose father had been a dignitary of Kayembe Mukulu, a chieftaincy with a strong tradition of independence from the Musuumb, soon made it clear that he intended to extend equal treatment to all ethnic groups within the new province. On November 18, 1963, Gaston Mushidi, who had undergone the ritual initiation at the Nkalaanyi since early September, was ceremonially confirmed and installed as Mwaant Yaav.[41] On that occasion Diur, who had brought with him a number of Cokwe and Ndembu ministers and legislators, first addressed the audience in French then, switching to Iruund, he added for the benefit of the "empire-minded" court:

> I was elected President of the Province of Lualaba. I am your chief and I will look after you. There are seven ethnic groups in the province. I am against favoritism and I will do my duty with full objectivity without taking any account of tribe or color.[42]

The new administrateur for Kapanga (Mbundj having now been promoted to the rank of district commissioner) had himself a few grievances to register about the new Mwaant Yaav. A Luba from North Katanga with no particular reverence for the Lunda paramount, he indignantly noted that Mushidi not only encouraged his subjects to ignore his authority, but also insisted that the administrateur and his staff must squat on the floor when drinking in his presence, abstain from smoking, etc. Moreover, he complained, "administration agents are summoned to appear before the traditional authorities. Every

problem, even those under the jurisdiction of the Territoire, is impera-
tively handled by the traditional authorities."[43] "This is unbelievable,"
the exasperated official wrote to Mushidi's secretary. "If need be, we
shall have to act like the colonialists, for the Adminstration is and
always will remain the same."[44]

This little war of attrition went on more or less intermittently
through the spring of 1964 while at Kolwezi President Diur tried to
walk a perilous tightrope between his own sense of a need for national
reconciliation and the contradictory pressures of those—whether
among the former "secessionists" or not—who tended to regard him as
Tshombe's creature. Tshombist forces in Lualaba, now inextricably
tied to the Lunda court and led by Thomas Kabwita, were especially
vociferous against what they viewed as Diur's "betrayal." The presence
of hundreds—or even thousands, according to some—of former
Katanga gendarmes roaming the bush with their equipment served as
a constant reminder of the fragility of the central government's control
over Tshombe's former stronghold while reports of other gendarmes,
supported by white mercenaries, training across the border in Angola
nurtured the hopes and fears centered around the prospect of
Tshombe's return.

The extraordinary assortment of maneuvers conducted more or less
simultaneously by the former secessionist leader during his twelve
months in exile would make the subject of an entire book, but is of
course of no direct concern to us.[45] As the political *chassé-croisé* continued in Léopoldville, armed insurrection deepened its hold over
Kwilu, only two hundred miles from the capital, and the thought of a
resort to Tshombe, which would have seemed unthinkable only a few
months earlier, now appealed to an increasingly wide range of politicians—if only because the wily Katangan had deliberately been
promising to every faction that approached him exactly what they
wanted to hear. Tshombe's brothers, Daniel and Thomas, dutifully
acted as liaison with the groups waiting in Angola, with Western embassies, and, naturally, in Katanga itself, with the provincial authorities whose benevolent neutrality would be essential if a military operation should after all prove necessary. It did not, of course, and on June
26, 1964, Tshombe returned to Léopoldville, just in time to witness
Adoula's resignation as prime minister and to be invited by President
Kasa-Vubu to form a new government.[46]

Tshombe's astonishing triumph was of course perceived—and

loudly celebrated — as a personal victory by his political associates in Lualaba, as well as by Lunda court circles. Diur's patient efforts to reconcile ethnic differences, which, only a month earlier, had led one report to predict that in the event of a Tshombist uprising, the Cokwe would "fight on the side of the Lunda,"[47] now threatened to be undone by a bout of Lunda chauvinism. Tshombe's kinsmen apparently viewed themselves as possessed of some viceregal mandate and gave a free rein to their disregard for local authorities. Four days after Tshombe's appointment as premier, Mushidi and Kabwita traveled to Angola without bothering to inform provincial officials, to supervise the return of the former Katangese gendarmes.[48] Within another week, according to a Léopoldville newspaper, eight thousand of these men had been gathered in Kolwezi alone.[49] A few weeks later, the same two chiefs, touring the southern regions of the Lunda heartland, summoned the administrateur for Dilolo (a Cokwe) to Chief Dumba's village and berated him vehemently for not putting his official powers at the service of the traditional chiefs, for showing favoritism toward the Cokwe, and for tolerating the activities of Angolan political refugees in his territory when Portuguese authorities had expressed their "unhappiness" about the situation.[50] As for the administrateur at Kapanga, whose lack of proper reverence had so incensed Gaston Mushidi, he was promptly replaced by one Mwasaza, a local landowner known for his loyalty to the Tshombe clan.

A new challenge to Lunda supremacy was confronted by the Mwaant Yaav in the eastern portion of Dilolo Territory where the predominantly Ndembu population, which had in the past identified almost totally with the northern Lunda and returned a solidly Conakat vote in 1960, was now being organized into a separate political-ethnic association known as the "Ndembo-Frères." While visiting Chief Sakayongo, Mushidi was lavishly entertained at Kisenge by the BCK-Manganèse mining company; then, addressing a mass rally at the local airfield, he launched into an acerbic diatribe against the Ndembu leader and assemblyman, Benjamin Bumba:

> There he is, this Bumba who would divide the children of a same father and mother, by saying "we are the Ndembo, those are Lunda," etc. Today, I ordered the dismissal of land chief M . . . and this Bumba too, when I get to Kolwezi, I will make my report to President Diur and he too will have to leave Kisenge; if he wants to carry on his politics, he will have to go to Léopoldville.[51]

Yet, if the Congo had become "one huge Katanga"—as both friends and foes of Tshombe often put it—it did not follow that Lunda would be given the same attention it had enjoyed during the secession. The new prime minister was far too busy planning the recapture of rebel-held territories with American, Belgian, and mercenary assistance to devote much time to the affairs of his own home area. In fact, the whole region that had been the stage of Tshombe's rise to fame was still left in the care of second-rank politicians (though under the eye of Munongo, now minister of the interior) and still partitioned into three provinces, despite Tshombe's earlier advocacy of a reunited Katanga.

Even after the tide of the rebellion had been turned back, the premier's attention was absorbed by the wholesale settlement, insisted upon by Belgium, of the intricate maze of financial claims and counterclaims between the Congo and the former colonial power (the so-called *contentieux*), and by the equally pressing problem of developing some semblance of a political base. The urgency of the latter issue was motivated by the constitutional obligation (which Tshombe himself had endorsed) of holding new general elections before March 30, 1965, at the latest. The task of organizing a genuine political party in a country traumatized by years of civil strife and still largely under military occupation was too chimerical to be even attempted, and Tshombe settled instead for a hastily improvised front collecting a rather disparate assortment of local "parties" and ethnic associations— forty-nine in all. This assemblage emerged from a two-week congress held at Luluabourg (February 7-20, 1965) and was given the rather unimaginative name of Conaco (Convention Nationale Congolaise) which, to many observers, sounded suspiciously similar to the name of Conakat.[52]

Conakat was indeed one of the constituent members of the new front, but in Lunda itself it now ran into an unexpectedly stiff opposition. Nationally, Conaco gained 122 out of 167 House seats (78 of them, admittedly, in former rebel areas that were still for the most part under military rule and where the local political class had often been liquidated), but in Lualaba province it won only two national seats against two also for Atcar, now led by relatively unknown men such as Oscar Mulelenu.[53] In the Lualaba provincial legislature, Conakat won nine of the fifteen seats, against five for Atcar and one for the "Ndembo-Frères" association, but this majority itself was split, it soon appeared, between a wing led by Dominique Diur and one led by Tshombe's brother, Thomas Kabwita, Chief Lumanga. In the Kapanga area, this cleavage ran essentially along the same covert lines which had separated the pro-Tshombe and pro-Ditende factions in 1960. Ernest Koji and Gabriel Kanyimbu Kambol, both of whom had been associated with the former Mwaant Yaav, were elected and rallied behind Diur while Oscar Mbundj was defeated. In Sandoa, a muted conflict of influence developed between Kabwita and Chief Mbako, son of the late Ditende.

The competition between Diur and Kabwita, which now came to center on the election to the governorship of Lualaba, not only reflected the continuing feud between the two royal lines (to which Diur himself was not a party), but also the opposition between the more intransigent Lunda "imperialists," now backed by the new Mwaant Yaav, and the more conciliatory position of Diur who, though a Lunda himself, had reached a working relationship with the Cokwe and Ndembu leaders. At this point, the issue of the disputed territory of Mwene-Ditu was suddenly injected into the Lualaba political imbroglio. This broader area had initially been claimed, it will be remembered, by two sections of the former Province of Kasai, but in November 1964, canceling an earlier decision, Godefroid Munongo had ordered that the referendum on the fate of Mwene-Ditu should offer a choice only between South Kasai and Lualaba. On the outcome of this referendum now hinged the question of whether the eight provincial legislators from Mwene-Ditu would take their seats in Mbuji-Mayi (the former Bakwanga) or in Kolwezi. As the lineaments of the Kabwita-Diur conflict emerged, the referendum was held on April 19, 1965, and produced an ostensible majority in favor of union with Lualaba.[54] The eight assemblymen from Mwene-Ditu now joined their fifteen colleagues from Lualaba and promptly sided with Thomas Kabwita, giving him — at least potentially — the needed majority to gain control of the provincial government. Diur's associates and the non-Lunda members protested this obvious maneuver and a protracted crisis ensued. On May 28 Munongo ordered Diur incarcerated on charges of misappropriating public funds, but even this form of intimidation did not break the deadlock.

For one thing, the outcome of the referendum was strongly objected to by Tshombe's own minister of agriculture, who was none other than Albert Kalonji, the former "Mulopwe" of South Kasai. South Kasai, as a matter of fact, had just returned the largest single group of Conaco representatives to the newly elected House (twelve), and its claims over Mwene-Ditu could hardly be ignored by the prime minister. Munongo, on the other hand, had been involved for weeks in a running controversy with President Kasa-Vubu over the issue of whether the Tshombe cabinet's mandate would automatically be terminated after the elections. Rather than an argument over constitutional law, of course, this dispute reflected Kasa-Vubu's growing fear that Tshombe might attempt to wrest the top executive post from him in the presidential elections scheduled to take place six months after the opening of the new parliament. Munongo's decision to seek the governorship of Katanga Oriental (prompted perhaps by a realization that his days as minister of the interior might be numbered) provided Kasa-Vubu with the needed excuse, and on the day after Munongo had been elected against his former secessionist colleague, J. B. Kibwe, he was dismissed from his cabinet post[55] and replaced by Victor Nendaka (technically a member of Conaco) without Tshombe's advice having been sought.

Meanwhile, the results of the Mwene-Ditu referendum had still not been ratified by the president, and Tshombe himself now admitted that as long as they had not, the eight assemblymen from that disputed area were not entitled to sit in the Lualaba provincial legislature.[56] Four days later, Dominique Diur was released from jail by the new minister of the interior and was promptly elected Governor of Lualaba against Thomas Kabwita with the backing of the Cokwe and Ndembu delegates.[57]

Kabwita's defeat in his bid for the governorship was also indirectly a defeat for the idea of Lunda supremacy in Lualaba. Ethnically and geographically the southwestern province came closest to representing a plausible receptacle for a modern version of the Lunda empire. Politically, too, the presence of Tshombe in the office of prime minister had seemed to augur well for the establishment of an indirect control of the province by circles tied to the Lunda aristocracy. Yet, like Muhunga's dreams for a "Cokwe province," the idea of a Lunda state, even an informal one, had now practically run its course and was slowly giving way to the notion of a polity in which no one ethnic group could hope to dominate its neighbors and where people who had indeed shared a common past (though not necessarily in the perspective cherished at the Musuumb) and continued to share in a common culture would have to develop new forms of coexistence and integration.

At the Musuumb, meanwhile, Diur's ascendancy was soon made perceptible by the appointment of a new administrateur de territoire. The unobtrusive Mwasaza was replaced by a "stranger" from the Luapula valley area,[58] but this appointment met with sullen defiance on the part of the Mwaant Yaav. A minor war of nerves now developed between the royal capital and the government post, five miles away. Where Mushidi had previously treated the agents of the provincial government as menials, he now went on to boycott them and to obstruct their action, forbidding them access to several parts of his chefferie and refusing to let them set up a field office at the Musuumb.[59] Such provocations were harmless enough by themselves, but the presumption that Tshombe's support would be forthcoming, should the situation degenerate into an open crisis, only served to encourage the Mwaant Yaav's arrogance.

In fact, though Mushidi may not have fully realized it at the time, Tshombe himself was now fighting for political survival. Ever since the elections, and particularly since the appointment of Victor Nendaka as minister of the interior, the prime minister had known that there would have to be a showdown with President Kasa-Vubu and with some of the politicians he had kept out of office for over a year. Paradoxically, the size of Tshombe's parliamentary majority represented a drawback as well as an asset for the prime minister, to the extent that it made it technically possible for him either to win the presidency or, by amending the constitution, to strip that office of most of its power, thus constituting a permanent threat to Kasa-Vubu's position. Other politicians — and the head of state himself, it would appear — also felt that Tshombe had outlived his political usefulness now that the rebellion's back had been broken, and that his controversial personality now stood in the way of a national reconciliation. The preliminary organizational session of the new parliament revealed that Tshombe's majority was far from monolithic and Victor Nendaka, breaking officially with the premier, attempted to organize the opposition in the House. But there was to be no confrontation for, in his opening address to the first regular meeting of the legislature on October 13, 1965, Kasa-Vubu simply terminated the mandate of the Tshombe cabinet.

The rest of the story is well known. Tshombe managed to block the election of Evariste Kimba (another of his former associates from the days of the secession) to the post of prime minister and, after the situation had become totally stalemated, the army high command deposed President Kasa-Vubu and took over the reins of the government.

Tshombe's professed satisfaction with the military takeover soon turned to bitterness when he discovered that some of his major adversaries, such as Nendaka and Bomboko, had been included in the new government while he was being kept on the sidelines. The thoughts of the former premier, out of office again for the second time in less than three years, seem to have turned almost automatically to the idea of another exile, but history now repeated itself even more literally when, as after Tshombe's earlier fall from power, the Mwaant Yaav died on December 12, 1965. Again there was the trip to the Musuumb, the anticipated election of another member of his family to the throne — this time his brother David Yav[60] — and then, three days before Christmas, the departure for Europe. Though he did not know it at the time, Tshombe would never see Lunda, or the Congo, again.

The epilogue of Tshombe's political career is too recent to require a detailed account. Though it ended in tragedy, not only for him but for hundreds of Congolese, Tshombe's second exile often seemed to verify the aphorism that history repeats itself as comedy. Paradoxically, while his legal position was technically stronger than it had been after the end of the secession (he was, after all, an elected member of the House and even his dismissal from office had never been sanctioned by parliament), his political stature was now much reduced, not only in the Congo where he was missed by few (Munongo, for one, had ostensibly rallied to the new regime) but also in Europe, where his past record seemed to generate embarrassment rather than confidence and where government and business circles had been quickly reconciled to the idea of prolonged military rule. Ian Colvin, a British journalist known for his sympathy toward Tshombe, called on the former premier in June 1966 and admitted:

> In his second exile he seemed rather to have lost touch with serious political friends and to be frequented by people of obscure background (attracted to him no doubt by the chance that he might return to a position of power), by arms dealers, ex-mercenary leaders and diplomatic agents. . . . He needed the close company of serious and responsible advisers, and these he seemed to lack. Belgium could find competent and devoted aides for whichever man was in

power. Africans in opposition were cast upon the European half-world of adventurers.[61]

Still, Mobutu's tentative moves toward a greater measure of economic nationalism, such as the canceling of mining and forestry concessions or the obligation for expatriate firms to relocate their headquarters in the Congo, created enough impatience with the new regime on the part of certain Belgian circles to blow some wind into Tshombe's sails and to lend credence to the ever-recurring rumors of a countercoup in his favor. Meanwhile, however, Mobutu had served notice on all would-be conspirators that he had no intention of letting himself be divested of power, by having four prominent congressmen (including Tshombe's luckless successor, Evariste Kimba) hanged publicly for their part in a poorly conceived plot to overthrow him. Tshombe, who had earlier been stripped of his parliamentary seat, was not involved in that particular attempt, but was nevertheless accused of treasonable activity, a charge that became all the more plausible when the Katangese "Baka" regiment of the Congolese National Army under Colonel Tshipola mutinied at Kisangani on July 23, 1966. From the beginning, the movement failed to rally the support of some of the mercenaries who had by now passed into the service of the new regime and after an uneasy stalemate during which Mobutu's prime minister, General Mulamba, and Godefroid Munongo both tried unsuccessfully to negotiate with the rebels, the mutiny was finally suppressed on September 25.

What was the situation in Katanga, and particularly in Lunda? There had obviously been continued contact between Tshombe and his brothers. As a matter of fact, Chief Lumanga had reportedly traveled to Madrid in March for that precise purpose. Whether such contacts involved treasonable plotting is of course another matter, but there was no doubt that the former premier enjoyed strong sympathies at the Lunda court, even though the new Mwaant Yaav carefully refrained from any inconsiderate moves. The case of Godefroid Munongo was more ambiguous. Munongo had officially endorsed the Mobutu regime and when, in April, the government had decided to merge Lualaba and Katanga Oriental into a single province of South Katanga, Munongo had been elected as governor over Dominique Diur, who had to content himself with the office of vice-governor. The antagonism between the two men probably explains why Diur and his associates now accused Munongo, Thomas Kabwita, and other Lunda chiefs of having plotted Tshombe's return to power, referring notably to the landing of a suspicious plane near the Mwaant Yaav's residence.[62] Intelligence reports indicating that some mercenaries had actually been recruited on Tshombe's behalf and the mid-September discovery by the French police of a small training camp in the Ardèche region probably helped persuade the Mobutu regime not to take any chances. Both Munongo and Diur were summoned to Kinshasa, then suspended, but while Diur was later cleared and allowed to pursue his official career outside Katanga, Munongo was eventually dismissed and interned on the island of Bula-Mbemba, where Gizenga had earlier been detained. As for Thomas Kabwita, Chief Lumanga, he chose to remove himself to Angola.

Tshombe and his brother Thomas were both tried in absentia in March 1967, along with the leaders of the Kisangani mutiny. Thomas, who was charged with having supplied Colonel Tshipola with a sum of one million Congolese francs to help finance the mutiny, was sentenced to a fifteen-year imprisonment, while Tshombe himself was sentenced to death. To the Lunda opponents of the Tshombe family, the time now seemed ripe for an attempt to unseat the new Mwaant Yaav (i.e., Tshombe's brother David Yav), who had not yet been ritually invested. In April 1967, charging among other things that David Yav's election had been irregular, Chiefs Mbako, Muteba, and Dumba persuaded the district commissioner to arrest the Mwaant Yaav, pending a full investigation. When he arrived at the Musuumb with a small military escort, however, the district commissioner was faced with a hostile crowd and with David Yav's adamant refusal to leave his capital — a move which, he understandably feared, would only be the prelude to his deposition. Thereupon, the district commissioner apparently lost his resolution and flew back to Kolwezi, leaving the troops to make their way back to Kapanga — which meant recrossing the entire length of the Lunda capital. As the crowd grew more threatening, the soldiers opened fire, killing three and wounding fourteen.[63]

It seems plausible that the central government, while not having directly ordered David Yav's arrest, was at least willing to recognize it as a fait accompli, had the district commissioner been able to carry out his original intention. Faced instead with a fumble, the government chose to let the matter rest as more urgent developments required its vigilance. Rumors of a new Tshombist uprising, heralded by a few incidents of sabotage perpetrated in Katanga, were mounting during the months of May and June. On June 30, however, the news that Tshombe had been kidnapped and was being held captive in Algeria (where he died two years later) left the plotters disconcerted and leaderless. Still, Katangese troops and white mercenaries launched a second mutiny at Kisangani on July 5. Simultaneously, smaller outbreaks flared up in the two eastern towns of Bukavu and Kindu. The mercenaries were soon forced to fall back upon the border city of Bukavu, which they managed to hold until the beginning of November, hoping for a diversionary attack by Tshombist forces from Angola.

Not with a bang, but with the proverbial whimper, the last sorry scene now came to an end—and fittingly enough it ended in Lunda. On November 1, a group of a few dozen ill-equipped and poorly led mercenaries entered the Congo from Angola at Luashi and briefly occupied Kasaji, Kisenge, and Mutshatsha. After a few desultory attempts to march on Kolwezi, then on Dilolo, the whole operation collapsed in the face of a swift reaction by the Congolese army, and within a week the mercenaries had withdrawn across the border for the last time. Tshisangama, the local chief, who had been appointed in 1963 by Gaston Mushidi and was known to be favorable to Tshombe, fled to Angola, but elsewhere in Lunda, whether through fear or apathy, not a single village had stirred.

For several months thereafter, contingents of the Congolese National Army were quartered in Lunda, keeping a watchful eye on the Musuumb and clearly awaiting the slightest excuse to intervene, but the Mwaant Yaav maintained a low profile and took the utmost care to avoid any incident. His moderation eventually paid off and in April 1968, one year after his attempted arrest, he was appointed administrateur de territoire for Kapanga.[64] This decision, coinciding with the Mobutu regime's gradual shift toward a more conservative policy line, represented a remarkable accolade for a man who had come so close to being deposed. At the same time, it eliminated the recurrent irritant inherent in the coexistence of two overlapping sources of authority over a same area. Familiar patterns of traditional rule now re-appeared as the Mwaant Yaav named a new Tshisangama, selected his sister to hold the dignity of Rukonkesh, and appointed

Gaston Mushidi's son Victor to replace Thomas Kabwita as Chief
Lumanga. In August 1968, David Yav was ritually invested and
assumed the title of Muteb II Mushid.*

Today, scattered around the royal enclosure as in a medieval
mystery play, the structures associated with each major protagonist of
recent Lunda history stand in mute testimony to the subtle interaction
of tradition and modernity. Here is Joseph Kapenda's modest brick
store flanked by the much larger and incongruously modern one built
by his ambitious sons. There are the houses of the major court dignitaries. There, staring at each other across a wide expanse of red
beaten earth, are the Methodist and Catholic churches. Down the road
are the graves of all the paramounts who reigned over Lunda since the
inception of colonial rule. The traces of Tshombe's tempestuous career
are elsewhere: in Lubumbashi, in Kisangani, in Belgium, Switzerland,
or Spain. Here, his trajectory is graphically foreshortened into a brief
walk of a hundred yards from his father's store to his brother's drab
little pillbox of a palace. Who is to say which is the truer perspective?

---

*Mwaant Yaav Muteb II Mushid died on November 24, 1973, at the age of 51, and was
promptly replaced by yet another brother of Moise Tshombe, Daniel, who thereupon adopted the
throne name of Mbumb Muteb. The new Mwaant Yaav was ritually invested in August 1974
(Communications from J. Jeffrey Hoover, 1 December 1973 and 15 December 1974).

9

Conclusion

Any attempt to assess the impact of alien rule on a traditional state such as Lunda is bound to be tentative at best, if only because of the complexity and multiplicity of the factors involved in what should ideally be a multilevel evaluation. There is, first of all, the problem of which time period to consider. The scope of this study was deliberately limited to the period of direct colonial rule and to its immediate aftermath, the first years of independence. Yet it would be overly simplistic to assume that the forces unleashed by European intrusions began to manifest themselves only when diplomatic convenience had pencilled in on a map of Africa which particular region would be allotted to which imperial power, or when the first European *boma* was set up in that region. Outside influences had been at work in Lunda long before the appearance of the first European soldier or trader, and it seems more than likely, as a matter of fact, that Lunda might not have achieved the particular form of organization and prosperity that it did without the stimulus of long-distance trade.[1] Yet it goes without saying that colonial rule, with all its deliberate emphasis on specific models of administrative "rationality" and economic "development," and with the more direct and forceful means of coercion it brought to bear on African societies, was a far more powerful factor of change than any other type of outside influence that Lunda had ever experienced since the days of Cibind Yiruung.

Another set of factors has to do with the social and political institutions of Lunda itself, and with the particular state in which they found themselves when colonial rule was imposed upon the area. Not all African societies seem to have had the capacity to absorb the sort of rapid, externally directed change which characterized colonial rule, nor did a superficially comparable acceptance of change necessarily result in parallel mutations. As a matter of fact, although Western observers have developed a number of intellectually satisfying yardsticks of "modernization" (by which is usually meant Westernization), the practical use of such criteria has often proved to be deceptive.[2] Finally, the input function of colonial rule is far from univocal. While the traditional emphasis on the contrasting policies and goals of the various imperial powers may have been exaggerated, the differential impact of a given colonial power's actions in the several parts of its empire can hardly be overstated. Geographical, climactic, and above all economic considerations can go a long way toward explaining how British colonial policy, though guided by a single ideology and implemented by the same breed of administrators, could produce such striking variations as could be observed, for instance, within the relatively narrow limits of a territory like Uganda — not to mention the obvious contrasts between northern and southern Nigeria or, in the case of the French, between Senegal and Mali.

Our understanding of the effects of colonial rule on a society such as Lunda also faces the unescapable fact that we perceive only imperfectly the nature of the Lunda political system. Some of the conceptual idiosyncrasies of its political culture (such as positional succession, perpetual kinship, or the plurilineal pattern of succession) are fairly easy to comprehend, but the total nature of the system evades our attempts at taxonomy. More important still — and just as elusive — is the question of whether Lunda political concepts themselves evolved as the fortunes of the "empire" waxed and waned.

Part of our problem comes from the fact that we do not know what to look for, or, conversely, that we know it too well. The enduring controversy over the use of the term "feudalism" in the context of precolonial Africa is a good example of this sort of perspective. Sheer ethnocentricity is probably responsible for most of the references to alleged "feudalism" in Africa on the part of early European travelers or administrators.[3] At a more sophisticated level of conceptualization, however, the reference to African "feudalists" by men like Delavignette reflected not so much a belief in the absolute parallelism between early medieval Europe and Africa, as the triumphalist attitude of Victorian Europe and the Jacobin-Weberian belief that centralization and "rational-legal" norms represented a step in the direction of greater individual freedom.[4]

More recently, much of the scholarly controversy over the applicability of a concept of feudalism to Africa (albeit an adjusted one) has involved the politically charged issue of the validity of Marxian analyses for non-Western societies, in much the same way as the equally heated dispute over the allegedly "classless" nature of African societies. The various and often conflicting definitions of African "feudalism" offered by authors such as Jacques Lombard, J.J. Maquet, or E.M. Chilvers (to mention only a few) are all more or less tailored to fit the specific societies which these anthropologists studied: the Bariba of Dahomey, the kingdom of Rwanda, or the interlacustrine states of East Africa. Lombard argues that

> in Black African societies, a "feudal" type system, or one with "feudal" tendencies offers at the same time an *unequal social structure* producing specific socio-economic or even political relationships, which might be called *vertical relationships,* between different social categories, and a *decentralized political structure* where regional governments enjoy some autonomy and economic privileges, but are linked to the higher central authority by personal ties which may be not only political, but also economic or kinship ties: these relations are called *horizontal.*[5]

There is no doubt that Lunda would fit this definition of African "feudalism," as well as many other more superficial ones. At the same time, however, Lombard and Maquet concur in viewing African "feudalism" not as a mode of production (as in the classical Marxian schemes); but as a political system of relations.[6] Lombard concludes, somewhat hastily, that the superimposition of European authority, inasmuch as it deprived the leaders of such a system of the very reason for their existence, was bound to be resented more deeply by "feudal" societies, because it wrecked the fundamental structures of these societies, released the subject groups, and deprived the aristocracy of their means of existence and traditional values. The successful adjustment to colonial overrule made by several "feudal" aristocracies would seem, at least superficially, to belie Lombard's analysis, but it is true that those political superstructures that offered the greatest apparent similarity to Western institutions past or present were usually the object of much greater attention — and thus interference — on the part of colonial authorities, if only because the Europeans felt that they could be presumed to function the same way as their supposed equivalents, and to carry the same socio-economic connotations.

The obvious fascination — whether policy-oriented or scholarly — felt by Europeans toward African state systems was reflected in the deliberate emphasis placed upon them by early essays in political anthropology, of which the best known, as well as the most valuable, remains Fortes and Evans-Pritchard's *African Political Systems*. A full generation and several reappraisals later, we are still indebted to them. With the fundamental contributions supplied by J. Middleton and D. Tait, Aidan Southall, M.G. Smith, and others, we now have a conceptual thesaurus which, while geared to Western mental yardsticks and thus fundamentally ethnocentric, has become as widely circulated among Africanists as Aristotle's categories among medieval Scholastics. Since the chief advantage of taxonomies is that they supply a set of commonly accepted definitions, I should not want to propose new or different conceptual referents, but simply to suggest, as others have, that the typologies of political organization developed by these writers should rightly be regarded as polar points, or as ideal paradigms, rather than as discrete, noncongruent categories.

Southall's now classical distinction between "pyramidal" and "hierarchical" authority systems[7] is particularly relevant to Lunda, especially if we bear in mind the fact that a model's adherence to reality should be judged in a diachronic perspective. Despite the many hierarchical features it acquired during the course of its relatively short history, Lunda had originally been a pyramidal system whose spectacular territorial expansion was predicated upon its ability to absorb segmentary or micropyramidal authority structures into its own larger pattern of vertical relationships. The ritual initiatory role of the fifteen Acubuung in the perpetuation of royal legitimacy reflects the original structure of the Lunda nucleus. We have noted that Lunda's remarkable capacity to extend its tributary network was based in large degree on the practice of leaving local land chiefs in position while incorporating them into a hierarchized fiscal structure. Where they existed, local dynasties were not infrequently left in power and have retained to this day their own self-contained institutions and procedures, notably in terms of succession to political office, involving the intervention of local electors parallel to those who preside over the Mwaant Yaav's ritual investiture.

Like Ashanti, with which it offers more than a few similarities,

Lunda has often been described as a "confederacy," and a popular though probably spurious etymology maintains that the name of Lunda itself was derived from a term meaning "friendship",[8] which would imply voluntary association. Like other originally pyramidal systems, Lunda acquired its earliest and most durable hierarchical features as a by-product of its military and fiscal pursuits. Specific military offices became institutionalized in the court hierarchy and specialized strategic duties were assigned to certain court dignitaries — a pattern reflected in the royal court's seating arrangement.[9]

It was the enforcement and collection of tribute, however, more than any other single factor, which contributed to the incipient transformation of Lunda from a pyramidal to a hierarchical authority structure. Tax collectors, royal inspectors, and resident "governors" were employed with increasing frequency, but recurrent instability at the center — born in part of the plurilineal succession pattern which constituted a permanent invitation to intrigue — prevented this structure from ever becoming fully stabilized. Indeed, while Lunda was remarkably successful in exporting its political culture, it was less effective when it came to institution-building. Authority roles remained for the most part functionally diffuse and procedural normalization was at best in an incipient stage of development.

Yet, the persistent trend toward permanent conversion from a pyramidal to a hierarchical system is unmistakable. Successive Mwaant Yaavs strove not only to keep the tribute flowing, but also to erode the position of local dynasties and to replace them with royal appointees. But the consistency of royal power was diluted by the conflictual features inherent in the mode of succession. Since all male descendants of past emperors, both patrilineal and matrilineal, were technically eligible, kingship was in fact elective and subject to protracted, often murderous rivalries in which factions recruited not only at the court but throughout the land regularly took part. Even when in possession of the throne, a reigning Mwaant Yaav would often find that *yikeezy* chiefs appointed by his predecessor remained loyal to the former king's lineage, of which themselves were often members, and that he was powerless to displace them.[10] Needless to say, satellite principalities which had preserved their own chiefly dynasties, such as the Sakambundj or the Kayembe Mukulu, could take advantage of these recurrent vagaries at the center to consolidate their local autonomy.

It is nevertheless conceivable that the fluctuating trend toward

greater centralization, still very much in evidence in the middle of the
last century, might have led Lunda to evolve in a direction similar to
that followed by Buganda a few decades earlier. Instead, the process
was brutally — and, as it turned out, permanently — interrupted by the
Cokwe "invasions" (or, to be more accurate, by a particularly bitter
sequence of succession feuds in which the Cokwe took an increasingly
central part), then by the establishment of colonial rule. All the same,
it is not at all obvious that Lunda could have achieved a greater degree
of concentration than it had reached over the previous two centuries.
Our knowledge of long-distance trade routes in south central Africa is
still quite incomplete, but there is little doubt (as shown by the
example of Kasanje) that, there as elsewhere,

> the development of long-distance trade seems to have been related
> to political development. For while political centralization may not
> have been indispensible for trade to develop and flourish, the de-
> velopment of trade itself in some areas favoured the creation of
> centralized political systems. And these in turn contributed in large
> measure to the further development of trade by providing organiza-
> tion and security for markets and caravans. A close link between
> long-distance trade and state organization must then be assumed in
> many states.[11]

One may speculate that the economic opportunities opened by the
establishment of the Portuguese on the Angola coast were a major,
perhaps even a crucial factor in the emergence of the Lunda state.
Certainly, Cinguud's deliberate efforts to establish contacts with the
Portuguese would indicate that commercial prospects were a key
determinant of Lunda's expansion. At the same time, the control of
trade routes and the ensuing prosperity it created were responsible for
the detachment of Kasanje, of the Kazembe and Bemba states, which
went into business for themselves, as it were, and soon became independent of the Lunda core. It might even be tempting to speculate
that the willingness shown by these states, which stood in no fear of
Lunda military intervention, to acknowledge their ritual and historical
ties with the Lunda nucleus was nothing more, in their eyes, than a
means to emphasize, in the face of those who might be tempted to supplant them as intermediaries, the fact that they enjoyed a privileged
relationship with the Lunda founts of trade.

During the second half of the nineteenth century, Lunda's economic position appears to have been challenged on several counts. Domestic resources that could be funnelled into the existing trade channels became scarce, especially ivory. At the same time, those regions of the interior which could supply Lunda with negotiable goods (roughly speaking, the area between the middle Kasai and Lualaba rivers) was penetrated by interlopers who not only displaced or circumvented existing intermediaries, but also undertook to organize the supply of their own trading activities. The successful state-builders of the late nineteenth century were no longer Lunda aristocrats but aliens like Msiri or ambitious local chieftains like Kalamba. Msiri drained the upper Lomami and Lualaba basins, and channelled its proceeds into the Nyamwezi and Arab trading network, thus undercutting the southeastern routes controlled by Kazembe and the Bemba, while Kalamba and his Cokwe associates established a monopolistic hold over trade goods from the middle Lulua basin. Not only were the trade routes to the coast once controlled by Lunda and its historical associates (Kasanje, Kazembe) jeopardized by these intruding groups, but also the Lunda state was cut off from its own economic hinterland by the outflanking effect of Msiri's and Kalamba's empire-building activities. By the end of the century, Lunda's decline was so pronounced that the chronic political instability caused by its seminormalized succession system now threatened its very survival.

In retrospect, it could be hypothesized that Lunda's failure to adjust to the changing conditions of long-distance trade may date back to the reign of Naweej II—the last period of continuity in Lunda history. Naweej's reported practice of allowing his trading partners to plunder designated portions of his empire in exchange for appropriating their goods was a tempting invitation for those traders to help themselves, but also revealed the court's unwillingness or inability to organize an effective fiscal or commercial monopoly in their own lands.[12] Furthermore, if we accept Robert F. Stevenson's compelling argument that there is a positive correlation between population density and state formation,[13] we can understand that Lunda's cohesiveness as a state may well have been marginal during most of its history. In this perspective, economic factors should be assigned a much larger role than is usually assumed in the rise and fall of the Lunda empire.

The major reason for attempting to reconstruct the nature of the Lunda state at the turn of the century is to enable us to form a clearer idea of the impact of colonial rule, and of the circumstances surrounding the establishment of a native administration system under Belgian control. One of the most common charges raised against Belgian rule in that area was that of having "dismembered" the Lunda empire or, in a somewhat related vein, of having deliberately rejected a policy of indirect rule which, in the eyes of officials like Duysters or Vermeulen, would have been ideally suited for the societies of western Katanga. Such charges were often restated after the war—and even after the Congo had become independent—without any further analysis of the historical record, and with a fairly clear realization on the part of those proffering them that what had been done could not, at such a late date, be undone.[14]

In actual fact, the circumstance surrounding the establishment of Belgian rule over the central portion of the former Lunda empire would suggest a rather different interpretation. There was of course no question of a protectorate treaty, or other bilateral agreement comparable to the Lewanika Treaty or the Buganda Agreement of 1900 between Lunda and the Free State. Except to support his claims on the diplomatic chessboard, Leopold was not in the habit of signing such documents with African chiefs, but even if he had been, he would have found it difficult (as had Dias de Carvalho) to find a responsible party with whom to negotiate such a relationship. The decrepit state of Lunda after two decades of Cokwe incursions cannot be doubted. As to the de facto alliance between Free State troops and Lunda elements against the Cokwe, it was of course a matter of tactical convenience, not a long-term policy commitment, yet it did not work to the disadvantage of Lunda, even though it benefited a particular faction of the aristocracy to the detriment of Mbumba's descendants.

There is, in any case, no reason to believe that Belgium's quarrel with Mushidi and Kawele might have stood in the way of a mutually beneficial arrangement with their successor any more than Lugard's forcible elimination of Kabaka Mwanga prevented the signing of the Buganda Agreement a few years later. As a matter of fact, as soon as the situation had been stabilized and Lunda had been transferred to Katanga, District Commissioner Gosme deliberately organized the native administration system around Muteba's still uncertain authority, and thereby restored Lunda control over areas which no Mwaant Yaav had effectively ruled for some thirty years. Even when official reversal of this earlier policy had led to the Kapanga Agreement of 1923, the Mwaant Yaav's paramountcy was still formally recognized over all but six chefferies of the Lunda heartland. At the same time, however, the Mwaant Yaav had been explicitly denied any direct executive authority outside of Kapanga territory and the post-1933 re-organization of the native administration system confirmed Lunda's reversion to a pyramidal authority structure. Still more important, the tributary network, a symbol of political allegiance rather than a truly fiscal apparatus, had been seriously jeopardized, and its continued operation made supererogatory by the colonial power's evident distaste for the principles on which it was based.

All the same, it seems unlikely that, given the disrupted condition of the Lunda state structure at the time of European occupation, the Mwaant Yaav's government could have wielded substantially broader powers under a different colonial system. Belgian authors and officials who wistfully speculated how Lunda might have fared under indirect rule seem to have labored under a number of commonly held misconceptions regarding the true nature of Lugard's system of native administration. British institutions have always held a strange fascination for continental Europeans, often for contradictory reasons, but nowhere was this more apparent than in colonial circles. The motley crowd of adventurers who constituted the vanguard of "civilization" in the heart of Africa had little enough regard for theoretical considerations, but as a more systematic form of rule was being organized, methods of native administration became the object of intellectual speculation, as witnessed by the transactions of the International Colonial Institute, and by the teaching dispensed at the French Ecole Coloniale or at the Antwerp Colonial Institute. Lugard's writings were highly esteemed among those circles, and even though the French still preferred to regard Lyautey as their patron saint, the policy of indirect rule came to be viewed by many officials as an ideal form of native administration.[15]

Much of this admiration was probably based on a rather superficial understanding of the actual operation of British field administration, but if the British themselves were mesmerized by Lugard's ideas to the point of applying them somewhat indiscriminately, one can understand that some Belgian officials may have attributed to indirect rule the same panacea-like virtues as those which the British parliamentary system held in the eyes of European political philosophers of the eighteenth and nineteenth centuries. Yet the policies rationalized by Lugard — and even more, perhaps, by some of his overenthusiastic disciples—in a theoretical form had originally been developed as an admirably pragmatic set of responses to a number of specific situations. In the words of Low and Pratt,

> Clearly, where [there] was a strong, extensive and centralized political system, indirect rule was not only possible but, given the limited power of the first British forces, enormously advantageous. In contrast, where the tribal system was diffuse, where it had been disrupted by earlier foreign contacts or where it lacked a central focus of power, then indirect rule did not suggest itself as a reasonable expedient or even as a possible policy.[16]

The supposed option between indirect and direct rule was, in many cases, not a choice at all. The record of British native administration in Tanganyika, in Nigeria, and elsewhere, as well as a close reading of Lugard himself belie the idyllic notion of a benign noninterference in local affairs by the colonial authorities. Even after we recognize that different types of traditional political systems did not lend themselves equally well to indirect rule, we have to take into further account the fact that all colonial officials tended to expect certain levels of performance from the chiefs, quite irrespective of their traditional legitimacy. Complaints of "corruption" or "inefficiency" on the part of European administrators seem to have been no less frequent in those areas where indirect rule was technically supposed to prevail, and the reaction was invariably one of direct, often heavy-handed intervention in the native administration process.[17] Ralph Austen concludes his study of northwest Tanzania under alien rule by suggesting that "conscious expressions of dedication to Indirect Rule (were) consistently accompanied by the strengthening of a bureaucratic apparatus which denie[d] the possibility of autonomous local development."[18] There is no valid reason to suppose that disarticulated Lunda would have been treated very differently from, say, the Haya kingdoms of Bukoba district had the Belgian authorities embraced indirect rule with the same degree of deliberateness as did Sir Donald Cameron in Tanganyika.

There was little or no deliberateness in the making of Belgian policy toward Lunda, unless one wishes to regard as deliberate the colonial administration's reluctance to restore those hierarchical features of the precolonial system which had lapsed for decades. The downgrading of tributary relationships, however, clearly derived from an ideological bias, just as the notion that no native administration unit should exceed the limits of a territoire proceeded from a rather narrow concept of administrative "rationality." The reversal to a pyramidal authority structure was reinforced after 1933, when all chiefs were subjected to the same legal status, and distinctions between the different categories of chiefs were increasingly disregarded. What remained of the Mwaant Yaav's adjudicatory role was also affected by this process. At the same time, however, his visibly reduced stature saved him from being considered by his former subjects in the Lunda heartland as a tool of the colonialists, leaving the institution of kingship imbued with a strong residual prestige.

European interference with the selection of Lunda paramounts was less spectacular than in many parts of colonial Africa. Although the Belgian authorities did not hesitate to demote even the most senior traditional chiefs (such as Kasongo Nyembo of the Baluba, or the Mwami of Kabare in Bushi), no Mwaant Yaav was, strictly speaking, deposed during the colonial period. Belgian interference confined itself to backing Muteba against Mushidi I after 1904, and thereafter to securing or endorsing the election of their favorite among the several eligible pretenders who were being considered, in accordance with custom, each time the succession was opened. Excepting Muteba's original installation, only two successions took place during the colonial period (in 1920 and 1951), a record of remarkable stability, whether by the standards of precolonial Lunda or by those of Belgian native administration policy. Although the Tshombe family often implied after 1951 that their descent from Mbumba and Mushidi had been held against them by the colonial administration, there is no evidence on record to suggest that this might have been a factor in Kaumba's election over Tshipao, or in Ditende's over Gaston Mushidi. There is, on the other hand, every indication that the Belgian authorities based their intervention in both cases not on the conflictual lines perceived by the Lunda aristocracy, of which they were only dimly aware, but on their own political criteria: Tshipao's claim was considered only to make Kaumba more tractable over the issue of "ethnic autonomy," and Ditende was backed against the unknown Gaston Mushidi on the basis of the ability he had demonstrated, during his thirty-three years of service as Chief Mbako, to assimilate European norms of performance. Neither Lunda custom nor Belgian policy favored direct lineal succession, and no single lineage was accordingly allowed to develop a proprietary claim on the throne.

In the vast arsenal of goods, ideas, norms, and stimuli showered by colonial rule over Africa, some operated more selectively than others. Not only was a group's capacity to absorb change a function of the intensity of the stimulations to which it was being subjected, or as David Apter argued, of the extent to which its material culture and values system were interdependent,[19] but the extent to which it felt capable of controlling and utilizing such innovations, as well as the expectation of specific benefits to be derived from change, proved to be of equal, if not greater significance. A crude but effective way of appraising the potential for lasting change inherent in a particular set of innovations would be to evaluate it in terms of what might be called its "opportunity component": to what extent was a given innovation perceived by its prospective recipients as an opportunity rather than as "an affair of the foreigners, which had to be endured"?[20]

In Lunda, the prevalent attitude seems to have inclined toward the latter view. From the beginnings of European occupation, colonial rule was associated with restrictions on long-distance trade, with heavy taxation, with road-building gangs into which the villagers were pressed, with labor recruitment, with forced cultivation of unfamiliar cash crops, with interference in the traditional processes, and only secondarily with a greater availability of consumer goods, with new economic and educational opportunities, with improved public health and tranquility. Clearly, the opportunity component of each particular set of innovations may be appraised differently under different circumstances: the road built at great hardship by reluctant villagers may eventually make it possible for them to offer their crops on more distant markets and bring medical services within closer reach, but too often, in Lunda as elsewhere, roads were built for administrative convenience and carried more touring officials, police platoons, and tax collectors than market-bound peasants.

Lunda's experience of the less constructive aspects of colonial rule was by no means unique and was directly related to its lack of exportable resources and to its remoteness from the industrial and commercial centers of the country. The historical core of Lunda, which already was in the process of being bypassed by the caravan trade during the latter part of the nineteenth century, now found itself more isolated than ever. Being more remote from the railroad and from the new centers of economic activity than any other section of the Lunda heartland, it was materially less affected by the new order than almost any other part of the former empire. During most of the colonial period — and even to this day — it was through the missions (practically speaking, the Methodist missions) that the Lunda court was introduced to Western goods, techniques, and values.

Although it could be culturally oppressive, this sort of contact was not as directly exploitative as those to which many other parts of colonial Africa were being exposed, but neither did it stimulate any radical departure from the self-contained type of economy on which Lunda had fallen back during its time of troubles. Yet the example of Joseph Kapenda, father of Moïse Tshombe and of the reigning Mwaant Yaav, shows that, given a proper incentive, the spirit of commercial entrepreneurship had not disappeared from Lunda. But the economic substratum of the Lunda state had been agrarian and the Mwaant Yaav's ultimate claim to paramountcry had rested on his title of Mwiin Mwaangaand (Master of all the lands). By contrast, the Cokwe (especially in the south, where Atcar had its strongest base) were more oriented toward trade, craftmanship and (later) salaried employment, and they tended to form a sort of rural proletariat. Land shortage was also a decisive factor among the southern Cokwe and Cokwe-ized communities, and communal tensions were clearly less pronounced where land was more plentiful or where alternative opportunities for economic gain were lacking, as in Sakundundu, Tshipao, or Kayembe Mukulu.

We have seen that Lunda-Cokwe relations had not initially been characterized by their hostility. The performance of specialized economic functions such as hunting or metal-working by the first Cokwe immigrants led to a relationship which has sometimes been described as a sort of symbiosis. But while the Lunda empire had found it possible to expand the territorial scope of its tributary network by extending a form of indirect rule over pre-existing sedentary communities, it has no adequate way of dealing with segmented groups of immigrants moving into politically organized provinces of the empire. So long as the flow of Cokwe migrants remained small, the easiest way of handling them was to treat them as a tolerated, but still marginal group, like the Jews of medieval Europe or the Dyula communities of West Africa, or if they decided to settle as farmers, to subject them to the existing hierarchies in return for the granting of cultivable land. As the flow of migrants increased, however, this solution became progressively unworkable. It was not so much that the Cokwe questioned the Mwaant Yaav's authority over the land, or even that of the subordinate Lunda chiefs, but rather that as traders, hunters, and artisans, they ignored the agrarian pattern of the Lunda state which was irrelevant to their pursuits. When the disappearance of the caravan trade, the collapse of the ivory-rubber economy, and the influx of European manufactured goods had forced many of the Cokwe into a life of sedentary farming, a substantial number of them accepted the political consequences of this new situation and acknowledged the authority of local Lunda chiefs. Those Cokwe who had entered Lunda as predators or those who were able to make a living by smuggling ivory, rubber, or wax into Angola in return for the lower-priced goods sold by the Portuguese during the early years of European occupation naturally rejected this type of accommodation and were eventually organized under their own chiefs.

The emergence of Cokwe territorial chiefs in the Lunda heartland may be viewed in retrospect as the recognition of a fait accompli, but it undoubtedly introduced a new dimension into the dynamics of Lunda-Cokwe relationships. When even the semblance of a tributary link with the Musuumb had been abandoned, it appeared that every Cokwe community in the heartland would sooner or later be viewed — or view itself — as an alien body. The fact that this did not happen until a generation later testifies to the extent of interaction (notably intermarriage) that had taken place between the two communities, as well as to the *Gesellschaft* base of Lunda national identity.

Conscious expressions of Lunda "nationalism" were few until the 1950s and seem to have arisen largely in response to the particularistic claims of the Cokwe, Lwena, and others. They were characteristically promoted by the traditional establishment (or by their educated sons) and stressed the inalienability of the ancestral land and the institutional nexus, rather than the ascriptive criteria of culture, language, or descent. It was, in that sense, a Hapsburg type of patriotism, while Cokwe solidarity was reminiscent of the Slavic national movements of the former Austro-Hungarian empire. It was also an aristocratic movement, stressing duties rather than rights and emphasizing the concept of "empire" rather than any form of "national sovereignty." Even so, the prestige of traditional institutions was sufficient to ensure a substantial amount of acceptance for the imperial premise, even in some of the more heterogeneous portions of the heartland, while the ethnocentric discreteness of Cokwe nationalism placed an absolute limit on its mobilizing potential.

Although the nature of Lunda political culture prevented the development of a strictly ethnic party, it would not have been inconceivable for the Lunda establishment to organize the defense of its interests through the medium of a "patron" party comparable perhaps to Buganda's *Kabaka Yekka* or to northern Nigeria's NPC. That this did not in fact happen was due in part to cleavages within the Lunda aristocracy, but also to the fact that Lunda had not been encouraged to regard itself, in law or in fact, as distinct from the rest of the Congo. The fact that Lunda, unlike Rwanda, Burundi, or Lesotho, was only a part of a much larger entity (albeit a recent one), as well as the absence of a strong tradition of local government during the colonial period, meant that political energies were not confined to the traditional arena and encouraged African leaders to seek power outside of the customary framework. Just as his father had extended his business activities far beyond the confines of the Lunda heartland, an ambitious man like Tshombe soon found a much more stimulating — and rewarding — scope for his talents in provincial, then in national politics. From that moment on, Lunda would be merely a card in a much more complex game, or at least one with much higher stakes. Tshombe's new-found power could of course be used secondarily to restore his family to the throne of Lunda but, by the time this happened, the issue was clearly no longer central among his preoccupations.

In fact, Tshombe never looked upon himself as a Lunda tribal leader, nor was he regarded as such either by his associates or by his opponents. The Weberian term "patrimonial leadership" which, as Zolberg suggested,[21] now characterizes most African regimes may represent the most adequate description of his political style. The highly personalized nature of the loyalty linking Tshombe (and other Congolese leaders) to their respective retainers transcended ethnic affiliations and made them, in most cases, irrelevant. Jean-Claude Willame points out that

the formation of personal clientages and territorial estates in the Congo did not arise from any popular support organized on an ethnic basis. Congolese regional leaders and key figures in the central government were able to pursue their political goals without much interference from the groups they supposedly represented.[22]

We have noted that despite the traditional background of several of its members, the government of secessionist Katanga made no serious effort to encourage direct involvement by the chiefs in the political process. In confining the senior chiefs of Katanga to an honorific role, Tshombe and Munongo were, in effect, setting aside precolonial legitimacy as a foundation for their authority. Rather, they attempted to appropriate for themselves a neotraditional form of authority based on the perpetuation of the colonial order. In his study of the British legacy in northern Nigeria, Heussler observed that "many things will go on being done . . . consciously or not, simply because the British did them that way."[23] And Zolberg reminds us that

> tradition (in Africa) does not merely refer to pre-European times. Many political institutions created during the colonial period have become, in the eyes of living men, part of the natural order of things: district commissioners, provincial commissioners, commandants and governors are offices hallowed by time; the African occupants of these offices derive their authority from the fact that they are the legitimate successors of the original charismatic founders.[24]

This tendency for African elites to move into the leadership roles vacated by the imperial rulers has perpetuated the sense of futility and powerlessness which so often characterized rural communities during the colonial period. The process of secularization which turned chiefs into petty bureaucrats goes on, even where their enduring authority has been given some form of official recognition, as in Lunda. By appointing the Mwaant Yaav to the post of administrator for Kapanga territory, the government of President Mobutu may implicitly have acknowledged his continuing hold over his people's loyalties, but they also endowed him with a new, delegated form of legitimacy which clearly integrates him for the first time into a nontraditional hierarchy. In that sense, this appointment represents the final act in the gradual bureaucratization of the chieftaincy rather than a belated vindication of the precolonial order. A subtle, covert ambiguity now cloaks the Mwaant's Yaav's dual status: the chief himself cautiously cultivates the illusion that his new administrative powers have been subsumed under his traditional prerogatives—by discharging his new duties from the Musuumb, for example, rather than from the governmental office at nearby Kapanga—while in the eyes of the national Department of the Interior, his status as a civil servant subjects him to hierarchical compliance with any directives issued by national, regional or subregional authorities. Although an open conflict between the two roles which the Mwaant Yaav is now supposed to perform may never arise under the present incumbent, the issue could be joined when the chieftaincy becomes vacant. Will a traditionally elected paramount automatically inherit his predecessor's position in the civil service, or will the governmental authorities see to it that an administratively qualified appointee be elevated to the traditional dignity of Mwaant Yaav? In other words, will one role eclipse the other, and if so, will not the traditional office inevitably fade before its modern counterpart?

Only time can provide an answer to those questions, but in the meantime the task of reversing the sense of alienation bred by the sterility of "native administration" policies over three generations still remains to be faced. No tradition of a participatory form of local government subsists to ignite the mobilization of rural resources. The ideological vaccum of the colonial age is only beginning to be filled, in the most tentative way, by Mobutu's hortatory exertions in the name of *authenticité* but the tradition of paternalistic coercion is still too strong for a genuine sense of self-reliance to turn subjects into citizens.

Appendix
Notes
Bibliography
Index

Appendix

Principal populations and principal chiefs customarily subjected to the Mwata Yamvo[a]

Belgian Congo	Principal chiefs
Basanga and Balamba	Musaka, Kazembe Ketshila, Panda, Mufunga, Tondo, Pwetu, Mukumbi, Samaleng Tshipay, Kanampumba, etc.
Babemba	Kazembe ya Luapula, etc.
Makamba	Tshanyika, etc.
Ndembo	Musokantanda, Saluseke, Kazembe de Kasaji
Ampimin	Kayembe Mukulu, etc.
Bampende	Mwatakubanda, etc.
Kasongo Lunda	Kiamfu, etc.
Bakongo	Kwibanda Mutombo, etc.
Baminungu	Tshimbundu, etc.
Bene Kilo	Tshibaba, etc.
Luena	Katende, etc.
Amateng	Sakabundji, etc.
Bakete	Musevu, Kalengmany, etc.
Asalampasu	Bakawande, etc.

| Bena Kanjimbi | Mukaleng Mulolo, etc. |
| Akansangu | Tulume, etc. |

Angola

Tshokwe	Ndondji, etc.
	Mwatshisenge, etc.
Ayimbangala	Tshingudi,
	Kasaji, etc.
Luena	Nakatolo, etc.
Karanda	Mwine Randa, etc.
Kwinkundu	Kaungulo, etc.
Mukwaza	Tshibwika, etc.
Bampende	Mwatakubanda, etc.
Bena Tshishinga	Mwine Tshishinga, etc.
Baminungu	Mwa Tshibundu, etc.
Amakoz	Kabu Kantanda,
	Ntambwe, etc.
Tubanga	Mwaza Mutombo, etc.
Amashiji	Kapenda Kamulembe, etc.

Rhodesia (N.)

Babemba	Kalama Dijimba,
	Tshiti Mukulu,
	Mbwambu, etc.
Bakawonde	Mushima, etc.
Bakavula	Kazembe, etc.
Bene Kabanda	Kabanda, etc.
Bapalakasa	Tshiti Mukulu, etc.
Bakatshipuna	Mukaleng Nama, etc.
Luena	Tshinyama,
	Bakaluvalo,
	Bakabuzi, etc.
Lutshazi	Babunda, etc.
Ndembo	Ikenge Kalulu a Kajombo,
	Mwinelunga,
	Kakoma, etc.
Makamba	Shinda,
	Kanongesha,
	Nakaseya, etc.

aAppendix to letter of 25 February 1960 from Mwata Yamvo Yav a Nawej III, Emperor of the Lunda, to the Governor General of the Belgian Congo.

Notes

Preface

1. See List of Abbreviations.
2. William B. Cohen, *Rulers of Empire: The French Colonial Service in Africa* (Stanford, Calif., Stanford University Press, 1971), p. 79.
3. The kingdoms of Rwanda and Burundi were placed under Belgian administration in 1920, but the area was legally and administratively distinct from the Congo and "native administration" was covered by a separate body of legislation which came much closer to the British version of indirect rule than any corresponding norms applying in the Congo.
4. See for instance the identification of "subject" groups by Mwaant Yaav Ditende Yaav a Naweej III in a letter of February 25, 1960, to the Governor general of the Belgian Congo (reprinted here in Appendix).
5. P. Montenez, "Note sur l'identité coutumière des indigènes d'origine Lunda," *Bull. des Juridictions Indigènes et du Droit Coutumier Congolais*, 4:11 (1936), p. 270.
6. M. McCulloch, *The Southern Lunda and Related Peoples* (London, International African Institute, 1951), p. 5.
7. See for example O. Boone, *Carte Ethnique du Congo, Quart Sud-Est* (Tervuren, MRAC, 1961), which derives much of its taxonomical information from administrative reports.
8. A similar definition is used by a young Zaïrese historian, Edouard Ndua, in his recent thesis, "L'installation des Tutshiokwe dans l'Empire Lunda" (Université Lovanium de Kinshasa, 1971), p. 4.

1. Central Diffusion and Political Expansion:
   The Growth and Decline of the Lunda State

1. A significant step in that direction has been accomplished with the publication of Jean-Luc Vellut's monograph, "Notes sur le Lunda et la frontière luso-africaine," *Etudes d'Histoire Africaine*, III (1972), pp. 61-166.

2. Jan Vansina, *Kingdoms of the Savanna* (Madison, Wis., University of Wisconsin Press, 1966), p. 97.

3. Daniel Biebuyck, "Fondements de l'organisation politique des Lunda du Mwaantayaav en territoire de Kapanga," *Zaïre*, XI:8 (1957), p. 813; see also, F. Crine, "Aspects politico-sociaux du système de tenure des terres des Luunda septentrionaux," *African Agrarian Systems*, ed. Daniel Biebuyck (London, Oxford University Press, 1963), p. 160.

4. Crine, "Aspects politico-sociaux," p. 165.

5. The plural form of *mwaantaangaand* is *antangaand*, but for the reader's convenience, the singular form alone will be used for this and other Ruund terms such as *yikeezy, cilool, ngaand, Mwaant*, etc.

6. Daniel Biebuyck, "Fondements," pp. 793-797.

7. Vansina, *Kingdoms of the Savanna*, p. 81.

8. See Ian Cunnison, "Perpetual Kinship: A Political Institution of the Luapula Peoples," *Human Problems in British Central Africa*, 20 (1956), pp. 28-48, and "History and Genealogy in a Conquest State," *American Anthropologist*, 59, no. 1 (1957), pp. 20-31.

9. The decision by Mwaant Yaav Naweej (ca. 1821-1853) to exact tribute in the form of ivory was clearly a response to the improved market conditions for that commodity, following the Portuguese decision to abolish the royal monopoly on ivory (1834).

10. Vellut, "Notes sur le Lunda," p. 83.

11. Vansina, *Kingdoms of the Savanna*, p. 82.

12. In the case of the Kazembe of the Luapula, it appears that such autonomy was achieved in all but name within one generation. See A. C. Pedroso Gamitto, *King Kazembe and the Marave, Chiva, Bisa, Bemba, Lunda and Other Peoples of Southern Africa* (Lisbon, Junta de Investigações de Ultramar, 1962); Mwata Kazembe XIV, *My Ancestors and My People,* (London, 1951).

13. Biebuyck, "Fondements," p. 798. Vansina, *Kingdoms of the Savanna*, p. 78, describes them as twins, while Biebuyck indicates that they are not.

14. David P. Henige, *The Chronology of Oral Tradition: Quest for a Chimera* (Oxford, Clarendon Press, 1974), p. 35. For conflicting versions, see Biebuyck, "Fondements," pp. 798-799; Dias de Carvalho, *Ethnographia e Historia tradicional dos povos da Lunda* (Lisbon, Imprensa Nacional, 1890), pp. 57-64; M. Van den Byvang, "Notice historique sur les Balunda," *Congo*, 1937, I:4, pp. 426-430; Léon Duysters, "Histoire des Aluunda," *Problèmes d'Afrique Centrale*, XII (1958), p. 81; A. E. Lerbak and D. Munuung, eds., *Ngand Yetu* (Elisabethville: Eglise Méthodiste, 1963); *Nsang Ja Aruund* (Musumba?, 1969?).

15. Dias de Carvalho, followed by Van den Byvang, regards Konde as Mwaaku's wife (which does not necessarily rule her out as a successor). A similar version is presented in some of the oral traditions collected at the Museo de Dundo (Angola) and quoted in M. L. Bastin, "Tshibinda Ilunga, héros civilisateur" (Brussels; Université Libre, 1966) (mimeographed thesis), App., p. xxxiv.

16. The first attempt at dating Cinguud's migration was made by Dias de Carvalho; see *Ethnographia e Historia*, p. 78. The leader of the Lunda migrants was reported to have visited one Dom Manuel, Governor of Luanda, whom Carvalho tentatively identified as D. Manuel Pereira Forjaz, who was governor from 1606 to 1609 (in fact, 1607-1611 according to more recent research). Vansina, in "La fondation du, Royaume de Kasanje," *Aequatoria*, XXV:2 (1962), pp. 45-62, notes that Cinguud died en route and that contact with the Portuguese was made by his nephew and successor, Kasanje, but otherwise follows Carvalho's chronology and suggests the date of 1610 for the encounter. More recently, however, David Birmingham has offered serious evidence to suggest that Kasanje reached the coast before 1576 and that the Lunda migrants had crossed the upper Kwanza before 1548; see "The Date and Significance of the Imbangala Invasion of Angola," *Journal of African History* VI:2 (1965), pp. 143-152. This view has in turn been criticized by Vansina in "More on the Invasion of Kongo and Angola by the Jaga and the Lunda," ibid., VII:3 (1966), pp. 421-429.

17. Van den Byvang, "Notice historique," p. 433.

18. C. M. N. White, "The Balovale People and Their Historical Background," *Rhodes-Livingstone Journal*, no. 8 (1949), pp. 33-36, and *An Outline of Luvale Social and Political Organization* (Manchester, Manchester University Press, 1960), pp. 43-48.

19. For Cokwe traditions, see M. L. Bastin, "Tshibinda Ilunga"; H. Baumann, *Lunda* (Berlin, Würfel, 1935), pp. 139-142; H. Capello and R. Ivens, *From Benguella to the Territory of Yacca* (London, 1882), vol. I, pp. 187-190; I. Struyf, "Kahemba: Envahisseurs Badjok et conquérants Balunda," *Zaïre*, II:4 (1948), pp. 367-375; C. M. N. White, "The Ethno-History of the Upper Zambezi," *African Studies*, XXI (1962), pp. 15-18; E. Ndua, "L'Installation des Tutshiokwe", pp. 36-48.

20. Dias de Carvalho, *Ethnographia e Historia*, pp. 91-112; M. Plancquaert, *Les Jaga et les Bayaka du Kwango* (Brussels, IRCB, 1932), pp. 70-99; Struyf, "Kahemba," pp. 356-380; H. Van Roy, "L'origine des Balunda du Kwango," *Aequatoria*, XXIV:4 (1961), pp. 136-141; Vansina, *Kingdoms of the Savanna*, pp. 92-95.

21. See E. Labrecque, "La tribu des Babemba," *Anthropos*, XXVIII (1933), pp. 633-648; Audrey I. Richards, "The Bemba of Northeastern Rhodesia," in *Seven Tribes of British Central Africa*, eds. E. Colson and M. Gluckman (Manchester, Manchester University Press, 1951), pp. 164-193.

22. *Contra:* Dias de Carvalho, *Ethnographia e Historia*, p. 538; Van den Byvang, "Notice historique," p. 437; Cokwe traditions collected at the Museo de Dundo, quoted in Bastin, "Tshibinda Ilunga", App., p. xxxv.

23. Yaav a Yiruung and Luseeng (also known as Luseeng Naweej) may have been two distinct rulers. *Ngand Yetu*, a collection of Ruund historical traditions collected under the auspices of Methodist missionaries, identifies two separate rulers between Cibind Yiruung and Yaav Naweej (see Table 1). The traditions collected at Kasanje by Schütt and by Capello and Ivens also refer to Rweej's two *sons* and successors, Yaav and Naweej. Vellut ("Notes sur

le Lunda," p. 66) suggests that these variations show "the hesitation of traditions when it comes to tying the reigning dynasty to the origins of the Lunda principalities." One might add that the concepts of positional succession and perpetual kinship render all attempts to establish biological descent somewhat futile.

24. The first mention of that title in a European document (where it was rendered as *Mataiiamvua*) appears in 1756 in Manuel Correia Leitão's "Viagem . . . as remotas partes de Cassange", ed. G. Sousa Dias, *Boletim da Sociedade de Geografia de Lisboa,* 56a. ser. (1938), quoted in Vellut, "Notes sur le Lunda," p. 76.

25. Van den Byvang identifies them as Kaniok ("Notice historique," p. 549) while Duysters ("Histoire des Aluunda," p. 87) calls them Akawamba (i.e., Bakete) but indicates their location in a way that suggests that they were Salampasu. This is the view adopted by Vansina, *Kingdoms of the Savanna,* p. 83.

26. Duysters, "Histoire des Aluunda," p. 87. Van den Byvang ("Notice historique," p. 554) describes him as a son of Rweej.

27. Muteba, Mulaji, Mbala, and Mukanza, according to Duysters. Van den Byvang, following Dias de Carvalho, lists them in a slightly different order (see Table 1). Vansina, for his part, gives two different listings (*Kingdoms of the Savanna,* pp. 161, 254).

28. For the Kazembe of Luapula, see Mwata Kazembe, *My Ancestors and My People;* A. C. Pedroso Gamitto, *King Kazembe;* Ian Cunnison, "The Reigns of the Kazembes," *Northern Rhodesia Journal,* III:2 (1956), pp. 131-136, and "Kazembe's Charter," ibid., III:3 (1957), pp. 220-232; E. Labrecque, "Histoire des Mwata Kazembe, chefs Lunda du Luapula, 1700-1945"; *Lovania,* vols. XVI-XVIII (1949-1951). Two other chiefs also bear the title of Kazembe: one (Kazembe Mutanda), between the Kolwezi-Dilolo rail line and the Zambian border, rules over a Ndembu population, and the other one is his replica on the Zambian side of the border.

29. On these episodes see R. F. Burton (ed.) *Lacerda's Journey to Kazembe in 1798,* and its appendix, *Journey of the Pombeiros,* transl. by B. A. Beadle (London, Murray, 1873).

30. Joaquim Rodrigues Graça, "Descripção da viagem feita de Loanda com destino as cabeceiras do Rio Sena," *Annães do Conselho Ultramarino (Parte não oficial),* Series I (Feb. 1854-Dec. 1858), p. 134. Graça himself found a Portuguese trader by the name of Romão at the Musuumb when he arrived there. Two or three other Luso-African traders (Manuel Gomes Sampaio, Antonio Bonifacio Rodrigues, Francisco Pacheco) also visited the Musuumb at approximately the same time as Graça (Dias de Carvalho, *Ethnographia e Historia,* p. 577). For further details, see Vellut, "Notes sur le Lunda," pp. 115-139.

31. Trade was thus officially treated in the same perspective as tribute.

32. Dias de Carvalho, *Ethnographia e Historia,* p. 580.

33. According to the late Mwaant Yaav Muteb II Mushid (himself descended from Mbumba), this appointment had been meant to keep

Mbumba — who had claims on the throne — away from the capital (interview with Mwaant Yaav Muteb II Mushid, April 1972).

34. Vellut ("Notes sur le Lunda," p. 69) suggests that Lunda rules of succession (under which all descendants of present or former emperors were technically eligible to the throne) may have resulted in an accumulation of potential successors after the two long reigns of Naweej (1821-1853) and Muteba (1857-1873), thus accounting for the bitterness of succession feuds over the following years.

35. The murder of one of Mbala's pregnant wives was reportedly considered as an unprecedented act of cruelty and earned for Mbumba and his line the lasting resentment of a good portion of the aristocracy (interview with Mwaant Yaav Muteb II Mushid, April 1972).

36. See the description by Cdr. Michaux of his visit to Mushidi in 1896 (i.e., during the repression of the first episode of the so-called Batetela uprising, but before the Lunda counteroffensive): O. Michaux, *Carnets de Campagne* (Namur, Dupagne-Counet, 1913), pp. 352-359.

37. Van den Byvang, "Notice historique," pp. 207-208; Duysters, "Histoire des Aluunda," pp. 95-98; Hupperts, *Historique de la Chefferie du Mwate Yamvo*, Kapanga: April 4, 1917; E. Ndua, "L'Installation des Tutshokwe," pp. 92-112.

2. The European Penetration and Partition of Lunda

1. Birmingham, *Trade and Conflict in Angola* (Oxford, Clarendon Press, 1966), pp. 133-134. Lunda's commercial relations with Loango, as presented by Birmingham, are viewed with some skepticism by Vansina; see his review of *Trade and Conflict* in the *Journal of African History* VII (1967), p. 547.

2. Despite the interest shown by eighteenth-century Portuguese officials in the cross-shaped copper ingots (*croisettes*) and other copper artifacts issuing from Lunda and Kazembe, copper never seems to have represented a major export (see Vellut, "Notes sur le Lunda," p. 90).

3. From *pombo*, a slave fair, a name which was itself derived from *mpumbu*, one of the largest such marketplaces, situated southeast of Stanley Pool (now Malebo Pool).

4. Birmingham, *Trade and Conflict*, p. 136.

5. Dias de Carvalho, *O Jagado de Cassange* (Lisboa, 1898), p. 106.

6. Dias de Carvalho, *Descripção da viagem a Mussumba do Muatianvua* (Lisbon, Imprensa Nacional, 1890-1894), I, p. 273.

7. Birmingham, *Trade and Conflict*, p. 146.

8. José de Lacerda, *Exame das viagens do Dr. Livingstone* (Lisbon, 1867), pp. 337, 367; Burton, *Lacerda's Journey*.

9. "Journey of the Pombeiros," App. to *Lacerda's Journey*. See also A. Verbeken and M. Walraet, *La première traversée du Katanga en 1806* (Brussels, IRCB, 1953). On Honorato da Costa, see Vellut, "Notes sur le Lunda," pp. 99-110. The two *pombeiros* are usually described as mulattoes,

but they may well have been assimilated Africans (see Vellut, ibid., note 93).

10. Lacerda, *Exame das viagens,* pp. 99-100.

11. J. Rodrigues Graça, "Descripção da viagem," pp. 143-144.

12. On Kasanje's conflicts with the Portuguese during that period, see Otto Schutt, *Reisen im Südwestlichen Becken des Congo* (Berlin, D. Reimer, 1881), pp. 80-82.

13. Vansina, *Kingdoms of the Savanna,* p. 203.

14. For figures on the Angola slave trade from the sixteenth to the eighteenth century, see Birmingham, *Trade and Conflict,* pp. 40-41, 78-88, 113-115, 137, 141, 154-155.

15. E. da Silva Correa, *Historia de Angola* (Lisbon, Editorial Atica, 1937), quoted by Vansina, *Kingdoms of the Savanna,* p. 185.

16. Figures for ivory and wax are from Allen Isaacman, "An Economic History of Angola" (unpubl. diss. University of Wisconsin, 1966), quoted in Joseph C. Miller, *Cokwe Expansion, 1850-1900* (Madison, Wis., University of Wisconsin, African Studies Program, 1969), p. 34.

17. Dias de Carvalho, *Descripção da viagem,* I, p. 271.

18. Dias de Carvalho, *O Lubuco* (Lisbon, 1889), pp. 9-10.

19. A. van Zandijcke, "Note historique sur les origines de Luluabourg (Malandi)," *Zaïre,* VI:3 (1952), pp. 227-249.

20. Paul Pogge, *Im Reiche des Muata Jamwo* (Berlin, D. Reimer, 1880), p. 237. "The Lunda is a poor hunter or fisherman, he does not practice beekeeping as in Kioko. [He] does not understand rubber collecting. . . . He practices trade primarily to acquire articles with which to adorn his body."

21. Ibid., p. 77.

22. Miller, *Cokwe Expansion,* p. 7.

23. Ibid., pp. 14-15. The concept of "pawnship" is borrowed from Mary Douglas, "Matriliny and Pawnship in Central Africa," *Africa,* XXXIV:4 (1964), pp. 301-313.

24. On the "capitalistic" character of cattle in pastoral economies, see J. J. Maquet, *Afrique: les civilisations noires* (Paris, Horizons de France, 1960), p. 151-152.

25. The hypothesis of demographic pressure advanced by Miller in *Cokwe Expansion,* and in his "Cokwe Trade and Conquest in the 19th Century," in *Pre-Colonial African Trade,* ed. R. Gray and D. Birmingham (London, Oxford University Press, 1970, p. 182), is questioned by Vellut ("Notes sur le Lunda," p. 149), on the basis of Ladislas Magyar's account of the 1850s. These two views are not entirely irreconcilable, however, since overpopulation is very much a function of the availability of cultivable land, and the Cokwe heartland was notoriously infertile.

26. According to Pogge, *Im Reiche des Muata Jamwo,* p. 238, the blacksmiths of Musumba were mostly Cokwe. Vansina notes in *Kingdoms of the Savanna,* p. 219, that the Cokwe "never infiltrated into the areas of the Lwena, Luchazi and Mbunda, because there the inhabitants were also hunters."

27. Dias de Carvalho, *Ethnographia e Historia,* pp. 556-557.

28. Mbala's flight from Mbumba's forces was recorded by V. L. Cameron, who was at that time traveling through the northeastern part of the Lunda homeland.

29. For example, the Salampasu, who defeated and repulsed the Cokwe in 1884, just as they had beaten back Lunda encroachments in previous generations.

30. O. Schütt, *Reisen im Südwestlichen Becken des Congo,* pp. 150-164.

31. For the following section, see Hermann Wissmann, *Unter Deutscher Flagge* (Berlin, Walter & Apolant, 1889).

32. Note the German spelling, later changed to "Luluabourg." In view of the origin of the members of Wissmann's caravan, the local people elected to call it, more simply and more accurately, "Malanji" or "Malandi." Van Zandijcke, "Note historique," p. 242.

33. Wissmann, then and later, viewed himself as a servant of German imperial interests. When he had to hand over Luluaburg to agents of the newly created Free State, he reportedly informed the Bashilange that they had been turned over to a much lesser people who, as he scornfully put it, "only have one pair of pants where we Germans have two." Van Zandijcke, "Note historique," p. 246.

34. P. Jentgen, *Les frontières du Congo Belge* (Brussels, IRCB, 1952), p. 12.

35. These views continued to be held by German officials even after Bismarck had left office. See the opinion of Secretary of State Adolf Marschall, quoted in J. Willequet, *Le Congo Belge et la Weltpolitik* (Brussels, Presses Universitaires de Bruxelles, 1962), p. 15.

36. The first journey is described in H. B. Capelo and R. Ivens, *From Benguella to the Territory of Yacca* (London, 1882) and the second in Capelo and Ivens, *De Angola a contra-costa* (Lisbon, Imprensa Nacional, 1886).

37. Dias de Carvalho, *Descripção,* II, p. 326; *A Lunda ou Os Estados do Muatianvua* (Lisbon, Adolpho Modesto, 1890), pp. 28-94.

38. Dias de Carvalho, *A Lunda,* pp. 199-203.

39. Ibid., pp. 225-229. The treaty with Cisenge was full of restrictive clauses, obviously inserted at his request, whereby the Mwaant Yaav was denied any authority over the Cokwe while Portugal committed itself to keeping open the trade routes to the Kwango, to the Musumba, and to the Lulua.

40. Miller, *Cokwe Expansion,* pp. 63-65, suggests that the Cokwe who were crossing the upper Kasai at that time at the invitation of rival Lunda nobles represented a new generation of leaders who acted independently of their predecessors.

41. Dias de Carvalho, *Descripção,* IV, pp. 266-287; *Ethnographia e Historia,* pp. 661-665; *A Lunda,* pp. 283-339.

42. C. S. Latrobe Bateman, *The First Ascent of the Kasai* (London, G. Phillip & Son, 1889), pp. 104-105.

43. E. Janssens and A. Cateaux, *Les Belges au Congo* (Anvers, Van Hille-DeBacker, 1908-1912), I, p. 681.

44. A. J. Wauters, "Les prétentions portugaises sur le Mouata Yamvo," *Mouvement Géographique*, VII:22 (1890), p.91.

45. Quoted in P. Van Zuylen, *L'Echiquier congolais ou le secret du Roi* (Brussels, Dessart, 1959), p. 218.

46. M. Planquaert, *Les Jaga et les Bayaka du Kwango*, pp. 124-142.

47. Eduardo dos Santos, *A Questão da Lunda* (Lisbon, Agência Geral do Ultramar, 1966), pp. 174-175. Portuguese authorities, aware of the fluid political situation in the Lunda heartland, had apparently decided that the treaties signed by Dias de Carvalho might have to be renogotiated if and when political stability was restored.

48. Van Zuylen, *L'Echiquier congolais*, pp. 187, 219-220.

49. Quoted in Van Zuylen, *L'Echiquier congolais*, p. 219 and note, p. 222.

50. Eduardo Dos Santos, *A Questão da Lunda*, pp. 174-179 and Apps. 11 and 14.

51. Ibid., pp. 179-187 and Apps. 15-20. The day before he met Sarmento, Dhanis had noted in his diary: "In all frankness, I must admit that all this land is really under Portuguese influence" (*Journal de Route de F. Dhanis*, September 10, 1890; MRAC Archives; quoted in Vellut, "Notes sur le Lunda," p. 123).

52. Van Zuylen, *L'Echiquier congolais*, p. 223; Dos Santos, *A Questão da Lunda*, p. 156; Charles Liebrechts, *Léopold II, fondateur d'empire* (Brussels, Office de Publicité, 1932), pp. 179 ff.; Réne J. Cornet, *Le Katanga avant les Belges et l'expédition Bia-Francqui-Cornet* (Brussels, L. Cuypers, 1946), pp. 292-293.

53. The treaty was ratified on July 4, 1891 , by the Portuguese Chamber of Deputies and on July 7 by the House of Peers. No parliamentary ratification was of course required of the Free State since Leopold was its absolute monarch. The actual demarcation of boundaries on the ground took place between 1891 and 1894. Further adjustments, necessitated by inadequate knowledge of the hydrography of the Kasai at the time of the original agreement, were arrived at in 1910. Finally, by the convention of July 22, 1927, Belguim surrendered to Portugal a 1,400-square-mile area on the right bank of the Kasai at the southwesternmost tip of Katanga (the so-called Dilolo boot) in exchange for a strategically located postage-stamp territory, 1.2 square miles in area, the main advantage of which was the considerable amount of money its acquisition by Belgium would save a private firm during the relocation of the lower Congo railroad. On these various points, see P. Jentgen, *Les Frontières du Congo Belge*, pp. 53-61, 74-78; Dos Santos, *A Questão da Lunda*, pp. 235-273 and App. 26; Van Zuylen, *L'Echiquier congolais*, pp. 230-232, 373-380, 479-484.

3. The Consolidation of Belgian Rule: Initial Patterns of "Native Administration"

1. For these and related developments, see Réne J. Cornet, *Le Katanga*; A. Verbeken, *Msiri* (Brussels, L. Cuypers, 1956); James A. Moloney, *With*

Captain Stairs to Katanga (London, S. Low, Marston & Co., 1893); Frederick S. Arnot, *Bihé and Garenganze* (London, Hawkins, 1893).

2. Henri Delvaux, *L'Occupation du Katanga, 1891-1900* (Elisabethville, Imbelco, 1950), p. 33.

3. See *Maisha ya Hamed bin Muhammed el Murjebi yaani Tippu Tib*, with a translation by W. H. Whiteley (Kampala, East African Literature Bureau, 1970).

4. See J. Ceulemans, *La Question arabe et le Congo* (Brussels, ARSC, 1959), pp. 323-345 and 361-364.

5. O. Michaux, *Carnets de campagne*, pp. 261-278.

6. See A. van Zandijcke, "La révolte de Luluabourg," *Zaïre*, IV:9 (1950); *La Force Publique de sa naissance à 1914* (Brussels, IRCB, 1952), chap. V.

7. *La Force Publique*, p. 530. After the great famine of 1910-1911 in Angola, Cokwe aggressiveness subsided almost totally.

8. E. Verdick wrote in 1903: "I am almost convinced that the Kiokos [Cokwe] that were mentioned to me are simply Bihenos" ("Rapport sur la fondation des postes du Sud," 17 August 1903; AMC:IRCB 507, p.7). Verdick seems to have been in error in that particular case, but the observation is nevertheless significant.

9. Lt. Scarambone, "Rapport politique sur la situation générale du poste," Katola, 21 September 1904, AMC: AI 1375.

10. Miller, *Cokwe Expansion*, p. 72.

11. Michaux, *Carnets*, p. 353. "Since he is a hereditary foe of the Kiokos and as I myself hate with all my soul and strength those bandits who have inflicted so much suffering on all the peoples of the southern part of the Free State, I was extremely anxious to establish contact with him so that we could band together against those predatory hordes."

12. Van den Byvang, "Notice," p. 207; Duysters, "Histoire des Aluunda," p. 95.

13. Michaux, *Carnets*, p. 354-358.

14. "Chez le Muata Yamvo." *Mouvement Géographique*, no. 40 (1897), col. 470.

15. Ndua, "L'Installation des Tutshokwe," p. 109.

16. Duysters, "Histoire des Aluunda," pp. 96-97.

17. Labrique, "Exposé de la guerre Lunda-Tshiokwe" (5 March 1922), AMC:AI 1618, p. 6.

18. E. Verdick, *Les Premiers jours au Katanga, 1890-1903* (Brussels, CSK, 1952), p. 164. See also P. Montenez, "Historique de l'organisation administrative de l'ex-District de la Lulua" (20 January 1933), AMC: Cornet 1D. 385. This is indirectly confirmed by a letter written on March 22, 1903, by Second Lieutenant Declercq in which he noted that "the whole area situated between the Kalagni and Luishi rivers was abandoned and still remains so today." Quoted by Ndua, "L'Installation des Tutshokwe", p. 110.

19. Verdick, *Les Premiers jours*, p. 168. The same reasons were invoked eighteen months later by District Commissioner Chenot to justify the abandonment of Katola: "Regarding its political usefulness, it is located between

the Lunda [and] Kioko tribes and has no effective action on either". (Letter of 17 November 1904 to the Governor general; AMC:AI 1375.)

20. Labrique, "Exposé de la guerre," p. 7. (This would be approximately 9°30' south latitude.)

21. "Rapport politique, District du Lualaba-Kasai, Novembre-Décembre 1905," digested for King Leopold by Baert in Letter no. 504 of 11 March 1906; AMC:IRCB 722. The area corresponds to the modern *chefferie* of Sakundundu, one of the four purely Cokwe chieftaincies subsequently recognized by the Belgian authorities.

22. District commissioner for Lulua to Administrateur de territoire for Kapanga, 30 October 1922 (ATK).

23. Ndua, "L'Installation des Tutshokwe," p. 111.

24. Verdick, *Les Premiers jours,* p. 168.

25. "Rapport de la Mission Géographique du Lac Dilolo" (16 March 1907), AMC: AE 268, p. 46.

26. Verdick, *Les Premiers jours,* p. 164. The official in question (nicknamed "Mbako" by the Lunda) would appear to be either Declercq or Gallaix.

27. Duysters, "Histoire des Aluunda," p. 97; Hupperts, *Historique.*

28. Verdick, *Les Premiers jours,* pp. 164, 171-172.

29. Letter of 28 February 1904, digested by Baert in "Courrier du Département de l'Intérieur," AMC: IRCB 722.

30. Scarambone, "Rapport politique" (21 Sept. 1904).

31. See Verdick's opinion of him in *Les Premiers jours,* pp. 170-171.

32. Scarambone, "Rapport politique."

33. Vice-Governor General Costermans to Secretary of state in Brussels, 19 December 1904, AMC: AI 1375.

34. *Bulletin Officiel de l'Etat Indépendant du Congo,* 1905, pp. 203-205.

35. *La Force Publique de sa naissance à 1914,* p. 380.

36. Rapport de la Mission Géographique du Lac Dilolo, pp. 47-48.

37. Scarambone, "Rapport politique"; District Commissioner Chenot to Governor general, 17 November 1904, AMC: AI 1375. See also Rapport politique, District du Lualaba-Kasai, May 1906, digested by Baert in "Courrier du Département de l'Intérieur," AMC: IRCB 722.

38. Letter of 23 June 1905, digested by Baert, in "Courrier du Département de l'Intérieur," AMC: IRCB 722.

39. *La Force Publique,* pp. 381-382; Montenez, *Historique.*

40. Miller, *Cokwe Expansion,* p. 73.

41. Gosme, *Notice sur les populations du District,* District de la Lulua, 29 June 1913, p. 8.

42. Governor general to Vice-governor general of Katanga, 1 April 1916.

43. A provincial ordinance of 26 June 1912 delineated the boundaries of the new District of the Lulua, in which Belgian Lunda was included, and established its administrative headquarters at Dilolo. The seat of the district was transferred to the newly created post of Kafakumba later that year (11 November 1912), then on 23 January 1915 to Sandoa, where it remained until

the district became part of the larger District of Lualaba in the 1933 reorganization.

44. District de la Lulua, Rapport annuel (1913), p. 7. That firm closed down its last posts in 1921. By that time the collection of rubber (and wax) in the district was "definitely terminated." Rapport politique (4th quarter, 1921), p. 1.

45. District de la Lulua, Rapport trimestriel (1st quarter, 1913), p. 2. See also District du Kasai, Rapport annuel (1913), AMC: AI 1373; letter of 16 July 1913 from the *Substitut du Procureur du Roi* to the *Procureur du Roi* at Elisabethville, relating how the branch manager of the Compagnie du Kasai at Dilolo had been held up by Cokwe tribesmen and forced to pay a "fine" for having traveled along a road which they had declared "off limits to Whites."

46. District de la Lulua, Rapport trimestriel (1st quarter, 1913), p. 2.

47. Ibid., pp. 3, 5.

48. District de la Lulua, Rapport trimestriel (2d. quarter, 1913), p. 5.

49. Ibid. (1st quarter, 1913), p. 11.

50. District de la Lulua, Rapport annuel (1913), p. 2.

51. District de la Lulua, Rapport trimestriel (2nd quarter, 1913), pp. 2-3.

52. Ibid. (1st quarter, 1914,), p. 3; (2nd quarter, 1914), p. 5.

53. *Bulletin Mensuel du Congo Belge*, 1915, p. 91.

54. Quoted in District de la Lulua, Rapport annuel (1923), p. 17.

55. District de la Lulua, Rapport annuel (1929), App., p. 4.

56. Inspecteur d'Etat (for Vice-governor general for Katanga) to Governor general, 6 June 1916.

57. Governor general [Henry] to Vice-governor general for Katanga, 1 April 1916.

58. District de la Lulua, Rapport trimestriel (2nd quarter, 1916), p. 6.

59. Territoire de Kapanga, Rapports politiques (2d, 3rd, and 4th quarters, 1917), passim (ATK).

60. Ibid. (3rd quarter, 1916), p. 2; (4th quarter, 1916), p. 2. (ATK).

61. District de la Lulua, Rapport annuel (1923), p. 17. Territoire de Kapanga, Rapport politique (4th quarter, 1916) (ATK).

62. Ibid.

63. District de la Lulua, Rapport sur l'administration générale (2nd half, 1924), p. 22. District du Lualaba, Rapport AIMO, Note synthétique du Commissaire de District (1934), pp. 19-20. However, another Lwena notable, Tshilemo, indicated his willingness to acknowledge the Mwaant Yaav's paramountcy — hoping perhaps to be placed in charge of the entire Lwena population in the Congo — but his followers numbered only 20% of that community.

64. L. Van Not, *Notice concernant les Lunda et les droits du Mwata-Yamvo sur des Bakete habitant le District du Kasai* (Luisa, 14 June 1919), AMC: AI 1619. District de la Lulua, Rapport annuel sur l'administration générale (2nd and 3rd quarters, 1921); idem, Rapports politiques (1st and 2nd half, 1923); idem, Rapports semestriels sur l'administration générale (1st half, 1924; 1st and 2nd half, 1926). Territoire de la Kasangeshi, Rapport politique (1st half, 1923); Van den Byvang, "Situation en région Bakete et

limites territoriales d'avec le District du Kasai" (Sandoa, 1923); Letters of 5 June 1923 and 26 June 1923 from AT Kasangeshi to AT Bakete; Letter of 19 June 1923 from AT Bakete to AT Kasangeshi; Letter of 13 December 1923 from District commissioner for Kasai to District commissioner for Lulua and reply of 8 February 1924.

65. Such was, for example, the case of Muyeye (Acting District commissioner for Lulua to Administrateur de territoire for Kapanga, 14 September 1917) (ATK).

66. Territoire de Kapanga, Rapports politiques (2d quarter 1917, 1st quarter 1919, 1st half of 1924, passim) (ATK).

67. District commissioner for Lulua to Administrateur de territoire for Kapanga, 14 September 1917 (ATK).

68. Georges Brausch, *Belgian Administration in the Congo* (London, Oxford University Press, 1961), p. 43.

69. Louis Franck, "La politique indigène, le service territorial et les cheffieries," *Congo*, 1921, I, p. 201.

70. Ibid., p. 196. Franck did not exclude the exceptional use of "our Negro clerks or NCO's" to fill these positions (p. 197).

71. Brausch, *Belgian Administration*, p. 42.

72. Ibid., pp. 198-199. The expression bears a striking resemblance to the terms used to describe the native administration policy of another admirer of Lugard, Sir Donald Cameron, governor of Tanganyika: "In a first period, the most important task was to stop the disintegration process by removing its causes and by resuscitating the life systems of native life and discipline, then to revive the spirit of cohesion and to build a progressive, intelligent and self-respecting society." Sir Edgar Twining, Despatch from the Governor of Tanganyika, no. 88, of September 20, 1950, quoted in Col. no. 227 (London: H.M.S.O., 1951).

73. Province du Katanga. Note du Commissaire Général, no. 94 (18 November 1919), quoted in Letter of November 3, 1920 (Commissaire général of Katanga to District commissioner of Lulua). The context of the 1920 reference (see chapter 4) confirms that the note applied, inter alia, to Lunda.

74. District de la Lulua, Rapport trimestriel sur l'administration générale (1st quarter, 1920). Idem, Rapport politique (2nd half, 1923). Bombolongo's election was later described as "a coup by a minority opposed to any dependency upon the Mwata Yamvo" (Territoire du Lubilash, Rapport politique (1st half, 1923), p. 1.

4. Prelude to Dismemberment:
Administrative Nationality and the Policy of "Ethnic Autonomy"

1. Labrique, "Exposé de la guerre Lunda-Tshiokwe," p. 1.

2. District commissioner of Lulua to the Administrateur de territoire for Kapanga, 9 October 1920.

3. Ibid.

4. District commissioner of Lulua to the Vice-governor general for Katanga, 9 October 1920.

5. Commissaire général (pro Vice-governor general) to District commissioner for Lulua, 3 November 1920.

6. Quoted in District de la Lulua, Rapport sur l'administration générale (1st quarter, 1921), p. 16.

7. Ibid., p. 10.

8. District de la Lulua, Rapport sur l'administration générale (1920), p. 17.

9. District de la Lulua, Rapport sur l'administration générale (1st quarter, 1921), passim. Idem (1st half, 1924), p. 8.

10. District de la Lulua, Rapport sur l'administration générale (2nd quarter, 1921), pp. 8-10. Idem (2nd half, 1925), p. 15.

11. The Belgian administration, typically, solved that problem by recognizing both contenders: Mwanza was put in charge of the eastern portion of the chefferie, and Kalyata of the western portion. The situation persisted until Mwanza's death in 1930.

12. The headman in question, Tshisenge Kapuma, was turned down by the Belgians and died in November 1921, but one of his nephews, Sakundundu, was eventually recognized as chief over one of the four autonomous Cokwe units created after 1923.

13. District de la Lulua, Rapport sur l'administration générale (1st quarter, 1921), pp. 8-9; idem (2nd quarter, 1921), pp. 3-5; idem (3rd quarter, 1921), p. 4.

14. District de la Lulua, Rapport sur l'administration générale (2nd quarter, 1921), pp. 24-25.

15. District de la Lulua, Rapport sur l'administration générale (3rd quarter, 1921), p. 2. In fact, a separate Tshipao unit has continued to exist to this day.

16. Territoire de Kapanga, Rapport politique (3rd quarter, 1920) (ATK).

17. Vice governor general to District commissioner for Lulua, 20 April 1922 (no. 1382/B4).

18. District de la Lulua, Rapport politique (2nd quarter, 1922), pp. 1-4.

19. Vice governor general to District commissioner for Lulua, 20 April 1922 and 12 May 1923.

20. District commissioner for Lulua to all Administrateurs de territoires (circular letter), 25 July 1922 (ATK).

21. Administrateur de territoire for Kapanga to District commissioner for Lulua, 1 October 1922 (ATK).

22. District commissioner for Lulua to Administrateur de territoire for Kapanga, 30 October 1922 (ATK).

23. Territoire des Tshiokwe, Rapport politique (1st half, 1923), p. 3 (ATD).

24. District de la Lulua, Rapport politique (2nd half, 1922), p. 3.

25. District commissioner for Lulua to all Administrateurs de territoires (circular letter), 11 May 1923 (ATK).

26. In 1929 the District commissioner estimated as follows the percentages of Cokwe living under the Lunda subchiefs: Muteba, 62.5%; Mbako, 57.7%; Lumanga, 48.7%; Mwene Ulamba, 70%; and Tshisangama, 75%. Com-

piled from District de la Lulua, Rapport sur la Situation Politique (1929), p. 4.

27. Territoire de la Kasangeshi (Kapanga), Rapport politique (1st half, 1923), p. 2.

28. MS by E. Wouters (1921), Dossier Mbako (ATS), quoted in Ndua, "L'Installation des Tutshokwe," p. 92.

29. Territoire de la Kasangeshi (Kapanga), Renseignements politiques: Contrôle de la déclaration de Bangu II (1 July 1921).

30. District de la Lulua, Rapport politique (1st half, 1923), pp. 27-28.

31. Territoire de Sandoa, Rapport politique (1st half, 1923), p. 2.

32. Territoire des Tshiokwe (Dilolo), Rapport politique (1st half, 1923), p. 3.

33. District commissioner for Lulua to Administrateur de territoire at Dilolo, 7 July 1923. Three months later, the District commissioner recognized that many of the Cokwe headmen he had met while on tour indicated their intention to go on living under the authority of a Lunda chief. (District commissioner for Lulua to Vice-governor general for Katanga, 14 October 1923.)

34. Territoire de Sandoa, Rapport politique (1st half, 1923), p. 3.

35. Appendix to: District commissioner for Lulua to Vice-governor general for Katanga, 14 October 1923, no. 1852.

36. District commissioner for Lulua to Vice-governor general for Katanga, 22 October 1923.

37. Idem, 14 October 1923.

38. Ibid.

39. Vice-governor general for Katanga to District commissioner for Lulua, 31 October 1924. See also Rapport aux Chambres Législatives sur l'administration du Congo Belge pour l'année 1925 (Brussels, 1927), p. 128: "Because of its periodicity, this royalty has assumed the nature of a political tribute in the eyes of the Batshiokwe. Hence a number of difficulties which we hope to eliminate through an agreement establishing a certain number of annuities, after which any dependency toward the Ulunda on the part of the Batshiokwe, even over land matters, shall be considered extinct."

40. Vice-governor general for Katanga to District commissioner for Lulua, 27 March 1926, 13 December 1926, 3 January 1931; District de la Lulua, Rapport politique (1931), p. 47.

41. District de la Lulua, Rapport sur l'administration générale (1st half, 1924), p. 8.

42. Ibid., p. 9.

43. District de la Lulua, Rapport sur l'administration générale (2nd half, 1924), pp. 1, 10.

44. Ibid. (1st half, 1925), p. 14; ibid. (1st half, 1926), p. 2; ibid. (1st half, 1927), p. 18. Chief Samazembe died in March 1931, but Samutoma had by that time become involved in a nativistic movement which an oversensitive administration viewed as "subversive," and he was arrested in January 1932. District de la Lulua, Rapport politique, 1931, pp. 61 ff.

45. Territoire de Sandoa, Rapport politique (2nd half, 1926; 2nd half,

1928); District commissioner for Lulua to Vice-governor general for Katanga, 13 October 1930, Appendix.

46. District de la Lulua, Rapport sur l'administration générale (2nd half, 1924), p. 18; ibid. (1st half, 1925), p. 2; ibid. (2nd half, 1925), p. 12.

47. District commissioner for Lulua to Vice-governor general for Katanga, 13 October 1930.

48. District de la Lulua, Rapport sur l'administration générale (2nd half, 1927), p. 1; ibid. (1st half, 1927), pp. 1, 22-23; ibid. (1st half, 1926), p. 1; ibid. (2nd half, 1925), p. 12; ibid. (2nd half, 1928), p. 6; Territoire de Malonga, Rapport d'enquête, Chefferie de Tshisenge (1935), MRAC.

49. M. Van den Byvang, "Organisation sociale chez les Aluunda" (1926), AMC: AI 1618, p. 1.

50. District de la Lulua, Rapport de sortie de charge (31 October 1926).

51. Vice-governor general for Katanga to District commissioner for Lulua, 13 December 1926.

52. District de la Lulua, Rapport sur l'administration générale (2nd half, 1926), p. 6.

53. District de la Lulua, Rapport politique (1931), p. 14.

54. Vice-governor general for Katanga to District commissioner for Lulua, 27 March 1926.

55. Ibid., 22 September 1926, p. 6.

56. Vice-governor general for Katanga to District commissioner for Lulua, 27 March 1926.

57. Idem to idem, 10 March 1925.

58. District de la Lulua, Rapport sur l'administration générale (2nd half, 1925), p. 2.

59. Vice-governor general for Katanga to District commissioner for Lulua, 22 September 1926. See also ibid., 27 March 1926.

60. District de la Lulua, Rapport sur l'administration générale (2nd half, 1926), p. 4; ibid. (1st half, 1927), p. 2.

61. Acting Commissaire général (pro Vice-governor general for Katanga) to District commissioner for Lulua, 28 October 1927.

62. District de la Lulua, Rapport sur l'administration générale (1st half, 1924), p. 4.

63. District de la Lulua, Rapport sur l'administration générale (2nd half, 1927), p. 27.

64. District de la Lulua, Rapport sur l'administration générale (1st half, 1924), p. 3; ibid. (2nd half, 1924), pp. 1, 13.

65. District de la Lulua, Rapport sur l'administration générale (1st half, 1924), p. 5; ibid. (1st half, 1925), p. 3.

66. District de la Lulua, Rapport sur l'administration générale (3rd quarter, 1920), passim; ibid. (2nd quarter, 1922), p. 4.

67. District de la Lulua, Rapport sur l'administration générale (1st half, 1926), p. 13. American Methodist missionaries and their African catechists were also accused of actively discouraging recruitment by the BTK. Territoire de Kapanga, Rapport politique, 1st quarter 1921 (ATK).

68. Territoire de Kapanga, Rapport sur l'administration générale (1st half, 1926), passim.

69. Ibid. (2nd half, 1926), pp. 3-4.

70. Ibid., p. 5. Emphasis in original.

71. District de la Lulua, Rapport sur l'administration générale (1st half, 1926), p. 12.

72. Vice-governor general for Katanga to District commissioner for Lulua, 22 September 1926.

73. Ibid.

74. Lord Hailey, *Native Administration in British African Territories* (London, H.M.S.O., 1950-53), vol. 4, pp. 12 and 20.

75. M. Crawford Young, *Politics in the Congo* (Princeton, N. J., Princeton University Press, 1965), p. 129.

76. Msgr. Stan de Vos, "La politique indigène et les missions catholiques," *Congo*, 1923, II:5, pp. 640-642.

77. L. Guébels, *Relation complète des travaux de la Commission Permanente pour la Protection des Indigènes* (Gembloux, Duculot, 1952?), pp. 249-250.

78. "Rapport au Roi de la Sous-Commission du Katanga pour la Protection des Indigènes," *Bulletin Officiel* (1925), no. 4.

79. Vice-governor general for Katanga to District commissioner for Lulua, 22 September 1926.

80. L. Duysters, Territoire de Kapanga, Rapport de Sortie de Charge (1 August 1927), AMC: AI(MT), II.5.2 (1409).

81. District commissioner for Lulua to Vice-governor general for Katanga, 16 August 1927, AMC: AI(MT), II.5.2 (1409).

82. Acting Commissaire general (pro Vice-governor general) to Governor general, 27 September 1927, AMC: AI(MT), II.5.2 (1409).

83. Vermeulen mentioned in a letter to the Vice-governor general for Katanga (12 October 1930) that Wamba, his former territory, has "80 organized chefferies and sous-chefferies, one non-traditional chefferie and some thirty unorganized chefferies."

84. District de la Lulua, Rapport sur l'administration générale (1st half, 1928), p. 2.

85. District commissioner for Lulua to Vice-governor general for Katanga, 31 March 1928.

86. District de la Lulua, Rapport sur l'administration générale (1st half, 1928), pp. 1-3; ibid. (2nd half, 1928), p. 3; ibid. (1929), pp. 3-5.

87. District de la Lulua, Rapport sur l'administration générale (2nd half, 1928), pp. 10-11.

88. Ibid., p. 11.

89. In 1934 the number of men recruited in Kapanga territory for long-distance employment had dropped to four.

90. See Vermeulen's own account of this program: "Les Villages-Modèles de la Lulua (1928-1934): Une expérience dans le domaine du bien-être indigène," *Revue Coloniale Belge,* no. 79 (January 15, 1949).

91. District commissioner for Kwango to District commissioner for Lulua, 10 September 1929.

92. J. Jammaers, Territoire de Kapanga, Rapport au sujet de la délégation "Lunda" envoyée au Kwango en mission politique, 9 April 1931.

93. Duysters, Rapport de sortie de charge, p. 3.

94. District commissioner for Lulua to Vice-governor general for Katanga, 12 October 1930.

95. Ibid., 13 October 1930, Appendix, p. 10.

96. Vice-governor general for Katanga to District commissioner for Lulua, 3 January 1931. A manuscript comment, apparently in the hand of the Vice-governor, on Vermeulen's letter of 12 October 1930, reads, "Mr. Vermeulen knows that there can be no question of reconstituting the Lunda empire."

97. Governor general to Vice-governor general, 10 October 1929.

98. Vice-governor general for Katanga to District commissioner for Lulua, 3 January 1931.

99. J. de Hemptinne, "Précisions sur le problème de la politique indigène," *Congo,* 1929, II:2, p. 197.

100. J. de Hemptinne, "La Politique indigène du gouvernement belge," *Congo,* 1928, II:3, p. 360.

101. Vice-Gouvernement Général du Katanga, Rapport annuel (1930), p. 8.

102. This was clearly understood by the administrator for Kapanga, who wrote in his 1931 report: "I have no desire to re-open the old polemic over the Lunda-Tshokwe-Aluena policy; this issue in which every one (of my predecessors) apparently got involved at the beginning of his incumbency always brought them in the end to the same result—namely that the Lunda, Tshokwe and Aluena must form completely independent groups." Territoire de Kapanga, Rapport politique (1931), p. 1 (ATK).

103. "If we permitted him [Kaumba] to travel through his twenty-two sous-chefferies," noted the Administrateur for Kapanga, "He would recognize that his Kingdom is still very important." Territoire de Kapanga, Rapport politique (1931) (ATK).

5. Stagnation and Change:
The Depression and the Crisis of Colonial Administration.

1. *L'Union Miniere du Haut-Katanga, 1906-1956* (Brussels, L. Cuypers, 1956), passim.

2. G. Van der Kerken, *La Politique Coloniale Belge* (Antwerp, Editions Zaïre, 1943), p. 143.

3. P. Piron and J. Devos, *Codes et Lois du Congo Belge* (Brussels, Larcier, 1959), II, 211, quoted in Young, *Politics in the Congo,* p. 131.

4. Alexandre Delcommune, *L'avenir du Congo Belge menacé* (Brussels, Lebègue, 1921), 2 vols.

5. By 1938 the Congo had become the largest single source of cotton for the Belgian textile industry, with a total of 37,671 metric tons, against 34,403

tons from the United States and 29,747 tons from India.

6. District de la Lulua, Rapport sur l'administration générale (1st quarter, 1921).

7. Ibid.

8. District of Lulua, Ordinance of 23 April 1925.

9. District de la Lulua, Rapport sur l'administration générale (1st half, 1928) p. 8.

10. Territoire de Kapanga, Rapports politiques, 2d and 4th quarters, 1918 (ATK), passim.

11. Idem, 1st and 3rd quarters, 1920 (ATK).

12. District de la Lulua, Rapport sur l'administration générale (2nd quarter, 1921).

13. J. Vandersmissen, "Essai d'introduction de materiel de culture à traction animale dans la région de Sandoa," *Bull. Agricole du Congo Belge,* December 1944, pp. 202-203, quoted in M. Miracle, *Agriculture in the Congo Basin* (Evanston, Ill., Northwestern University Press, 1967), pp. 273-274. Vandersmissen claims that cattle were common in Lunda when the first administrative posts were set up in the area in 1903. However, Edgar Verdick, who led the expedition which resulted in the founding of these posts, makes no mention of encountering cattle in Lunda, although he noted their existence farther to the north among the Luba and Kaniok. See "Rapport sur la fondation des postes du Sud," 1903, AMC: IRCB 507, and *Les premiers jours au Katanga.*

14. District de la Lulua, Rapport politique (1931), pp. 96-97.

15. District de la Lulua, Rapport annuel (1920), p. 5; District de la Lulua, Rapports trimestriels (1920), passim.

16. District de la Lulua, Rapport annuel (1923), p. 30.

17. District de la Lulua, Rapport politique (1931), p. 96.

18. Vice-governor general for Katanga to District commissioner for Lulua, 22 April 1932.

19. District de la Lulua, Rapport politique (1931), p. 96. Unnoticed at the time, one enterprising African trader, named Joseph Kapenda, had begun to lay the foundations of a small business empire by organizing the import of manioc from Kasai into Kapanga territory—hence the nickname of "Tshombe" (manioc) which he passed on to his elder son Moïse—or so legend has it.

20. District de la Lulua, Rapport sur l'administration générale (2nd half, 1928), p. 22.

21. Province du Katanga, Service des AIMO, Recensement de la population au 31 décembre 1930 (Elisabethville, 1931). Later still, in 1953, the proportion of foreigners in the African manpower of Elisabethville was still in excess of 20% (Ville d'Elisabethville, Service de la Population Noire, Rapport 1958).

22. The Union Minière, notably, recruited heavily outside of Katanga and, within the province, continued to recruit from sources other than the BTK/OCTK, whether indirectly through "freelance" recruiters or directly, after 1926, in the populous District of Lomami.

23. Rapport annuel de l'Office Central du Travail au Katanga, 1931 (Burssels, 10 March 1932).

24. Specifically from 12.31 months in 1920-1921 to 23.65 months in 1930. Rapport annuel de l'Office Central du Travail au Katanga, 1931.

25. These were based on the Pignet index, based on height, chest circumference, and weight.

26. Office Central du Travail au Katanga, Rapport annuel (1933), Appendix.

27. Ibid.

28. Office Central du Travail au Katanga, Rapport annuel (1931), p. 30.

29. Note pour les membres du Comité local de l'OCTK no. 10, 5 January 1931.

30. District de la Lulua, Rapport politique (1931), p. 104.

31. See H. Nicolai, *Le Kwilu: Etude géographique d'une région congolaise* (Brussels, Cemubac, 1963).

32. See Faustin Mulambu-Mvuluya, "Contribution à l'étude de la révolte des Bapende," *Cahiers du CEDAF,* no. 1 (1971), pp. 8, 25-27.

33. Ibid., p. 16.

34. Agent Sanitaire Drion to District commissioner for Lulua, 3 October 1931.

35. District de la Lulua, Rapport politique (1931), p. 11.

36. R. P. Supérieur, Mission des RR.PP. Franciscains de Dilolo, to Administrateur de territoire for Dilolo, 12 December 1931, quoted in District de la Lulua, Rapport politique (1931), p. 11; Territoire de Sandoa, Rapport sur la situation générale (1931), p. 2; Parquet de la Lulua, Rapport du Substitut J. Collignon, 26 December 1931, quoting from testimony of black catechists from the Franciscan mission at Sandoa.

37. Territoire de Sandoa, Rapport sur la situation générale (1931), p. 4; District de la Lulua, Rapport politique (1931), p. 20; A. T. Sandoa to District commissioner for Lualaba, 24 January 1940.

38. For this and related movements see E. Bustin, "Government Policy toward African Cult Movements: The Case of Katanga," in *African Dimensions: Essays in Honor of William O. Brown,* ed. Mark Karp (New York, Africana, 1975), pp. 113-135.

39. Province d'Elisabethville, Rapport AIMO (1934), p. 15.

40. District du Lualaba, Rapport AIMO (1937-1939), passim.

41. See Ian Cunnison, "A Watch Tower Assembly in Central Africa," *International Review of Missions,* XL, 160 (October 1951), pp. 456-468.

42. See Bustin, "Government Policy"; Dom Grégoire Coussement, "La secte du Punga et du Mama Okanga," *Bulletin des Juridictions Indigènes et du Droit Coutumier Congolais* (May 1935), pp. 64-67; Province d'Elisabethville, Rapport AIMO (1934), p. 13.

43. District du Lualaba, Rapport AIMO (1934), p. 24; ibid. (1936), p. 6; ibid. (1939), p. 4; ibid. (1943), p. 9.

44. E. De Jonghe, "Formation récente de sociétés secrètes au Congo Belge," *Africa,* IX, 1 (1936), pp. 56-63.

45. Without suggesting that this was necessarily significant, one can

observe that the main ingredient in *Ukanga* was water from the first rains, while *Tshimani* included vegetable substances (tree bark and oil) and *Tambwe* a variety of fragments obtained from humans and animals.

46. Bustin, "Government Policy," and references therein. The use of hypodermic needles may actually have appeared in conjunction with Ukanga, at least in the 1943 case of Chief Lumanga (interview with *Nswaan Mulopw* Mutshaila Ditende, April 1972.)

47. District de la Lulua, Rapport politique (1931), pp. 1-9; Province d'Elisabethville, Rapport AIMO (1934), pp. 1-4.

48. District de la Lualaba, Rapport AIMO (1934), p. 25.

49. Ibid., p. 6.

50. Administrateur de territoire for Malonga to District commissioner for Lualaba, 18 March 1934; District du Lualaba, Rapport AIMO (1934), pp. 41-42.

51. Ibid., p. 42.

52. Province d'Elisabethville, Rapport AIMO (1934), p. 75.

53. Minutes of the 1935 Conseil de Gouvernement meeting quoted in Letter from Provincial commissioner (interim) for Elisabethville to District commissioner for Lualaba, 15 October 1935.

54. These were, in Sandoa: Tshibamba, Muteba, Mbako, Lumanga, Tshipao, Sakasongo (Mwene Ulamba), and Kayembe Mukulu; in Malonga: Muyeye, Sakambundji, Saluseke, Mwene Kanduke (Dumba), Tshilemo, Sakayongo, Tshisangama, Kazembe, Mukonkoto, Mafunga, Kawewe, Mwene Kasanga, and Musokantanda.

55. District of Lualaba, Rapport AIMO (1936), p. 9. By the end of 1939 the total number of native administration units in the district had been reduced from 97 to 69 units.

56. District commissioner for Lualaba to Provincial commissioner at Elisabethville, 7 February 1936.

57. Provincial commissioner at Elisabethville to District commissioner for Lualaba, 9 March 1937.

58. Administrateur de territoire for Sandoa to District commissioner for Lualaba, 24 January 1940.

59. Administrateur de territoire for Malonga to District commissioner for Lualaba, 28 February 1940.

60. District commissioner for Lualaba to Provincial commissioner at Elisabethville, 5 April 1940. The rebates were discontinued after 1940. (Interview with Mwaant Yaav Muteb II Mushid, April 1972).

61. Pierre Ryckmans, *Etapes et Jalons* (Brussels, Larcier, 1946), p. 64.

62. District de la Lualaba, Rapport AIMO (1935), pp. 5-6. Administrateur de territoire for Kapanga to District commissioner for Lualaba, 30 May 1935 (ATK). The name of Chief Mbako (who actually succeeded to the throne in 1951) was suggested by the conspirators as a replacement for Kaumba.

63. Article 45. A decree of 16 February 1937 added reforestation to the already long list of compulsory tasks to be performed by villagers—without

remuneration, of course. On compulsory cultivation, see Mulambu-Mvuluya, "Cultures obligatoires et colonisation dans l'ex-Congo Belge," *Cahiers du CEDAF,* 1974, no. 6-7.

6. Out of the Depression and through the War: Bureaucratizing the Chieftaincy

1. Ryckmans, *Etapes et Jalons,* p. 28.
2. District du Lualaba, Rapport AIMO (1936), p. 21.
3. District du Lualaba, Rapport AIMO (1934), p. 68.
4. Compiled from data given in District du Lualaba, Rapport AIMO (1944) pp. 6-7.
5. Ibid., p. 8. This situation was by no means limited to Lualaba. In the Lodja area (Sankuru) the number of planters dropped from 20,000 in 1940 to 16,000 in 1943. In another area of Kasai, the number of planters declined from 20,737 in 1940 to 15,729 at the end of 1943. "According to COTONCO, the number of adults involved in the cotton crop in the Lomami region dropped from 58,000 in 1940 to 45,700 in 1943." Lomami-Kasai, January 1944, quoted in Antoine Rubbens, *Dettes de Guerre* (Elisabethville, Eds. de l'Essor du Congo, 1945), p. 45.
6. P. G., "L'administration des indigènes," *L'Essor du Congo,* 14 June 1945.
7. District du Lualaba, Rapport AIMO (1944), p. 10. The average agricultural output per planter for the entire province in 1943 was 314.7 kilograms, indicating that productivity in Lualaba was much higher than in other districts.
8. P. G., "L'administration des indigènes."
9. Ibid.
10. Dr. L. Mottoule, addressing the Elisabethville section of the University of Liége Alumni Association, *L'Essor du Congo,* 21-23 April 1945.
11. "Medical canvassing virtually disappeared during the war. Outside of the administrative centers, no more assistance was extended to the natives in this field" (District du Lualaba, Rapport AIMO, 1944, p. 13). "Only where antagonism between the different denominations stimulates a special effort do the children escape the mediocrity of the school system" (idem, 1945, p. 14).
12. The ratio of school attendance by children of the District of Lualaba declined steadily during the war, from 72% in 1940 to 61% in 1944.
13. The task of eradicating these movements was accordingly left for the most part to native jurisdictions. In Sandoa territory alone, native courts rendered 144 decisions against "diviners and sorcerers" between 1939 and 1944.
14. These rumors were ancient and possibly based on specific occurrences linked to the need for human victims in the performance of certain rituals. Over the years, a number of chiefs in Katanga and elsewhere were indicted for perpetrating or permitting human sacrifices in connection with their

ritual enthronement. In the 1920s, accusations of alleged kidnapping had been repeatedly directed against the Mwaant Yaav himself (see Bustin, "Government Policy").

15. Although they took on a new form during the war, rumors concerning an impending takeover of the Congo by "the Americans" had been in circulation long before. See for example District de la Lulua, Rapport politique (1931), p. 21.

16. Territoire de Kolwezi, Rapport AIMO (1945), p. 4.

17. Bustin, "Government Policy," pp. 116-117.

18. District du Lualaba, Rapport AIMO (1941), p. 2.

19. On the mutiny see "Mutineries au Congo Belge," *Zaïre*, XI, 5 (May 1957), pp. 487-514. On the abortive episode at Elisabethville, see Bruce S. Fetter, "The Luluabourg Mutiny in Elisabethville," *African Historical Studies*, II, 2 (1969), pp. 269-277.

20. District du Lualaba, Rapport AIMO (1944), p. 14.

21. According to the late Mwaant Yaav, Muteb II Mushid, Kaumba was suspected of being sympathetic to the Luluabourg mutineers. (Personal communication, April 1972.)

22. District du Lualaba, Rapport AIMO (1945), p. 3.

23. Rubbens, *Dettes de Guerre*, p. 48.

24. Ibid., pp. 46-47.

25. District du Lualaba, Rapport AIMO (1944), p. 18.

26. Rubbens, *Dettes de Guerre*, p. 192.

27. District du Lualaba, Rapport AIMO (1938), p. 6.

28. Provincial commissioner, Elisabethville, to District commissioner for Lualaba, 4 December 1934.

29. District du Lualaba, Rapports AIMO, 1935-1945, passim.

30. *Rapport sur l'administration du Congo Belge présenté aux Chambres Législatives*, 1949, p. 90.

31. "Although in the past we tried to keep track of every son of a notable, it seems that we should first of all seek out potential successors to watch their progress. Otherwise, we should have to keep track of too many pupils who will never be of any interest to the administration." Territoire de Kapanga, Rapport AIMO (1947), p. 8.

32. Territoire de Sandoa, Rapport AIMO (1945), p. 9.

33. Province d'Elisabethville, Rapport AIMO (1934), p. 111.

34. Territoire de Malonga, Rapport AIMO (1948), p. 11; Territoire de Dilolo, Rapports AIMO (1952), p. 34; ibid. (1953), p. 37; ibid. (1954), p. 40.

35. District du Lualaba, Rapport AIMO (1943), p. 9; Territoire de Sandoa, Rapport AIMO (1950), p. 34. Mutshaila Ditende was made *Nswaan Mulopw* in 1961 and still retained that position in 1973.

36. See Territoire de Sandoa, Rapport AIMO (1949), p. 7.

37. Territoire de Sandoa, Rapport AIMO (1946), p. 17.

38. Ibid., p. 15.

39. Territoire de Kapanga, Rapport AIMO (1947), p. 11.

40. Territoire de Kapanga, Rapport AIMO (1948), p. 9.

41. Ibid.

42. Ibid. (1949), p. 10. Kaumba was sixty-nine years old at the time.

43. Ibid. (1947), p. 7.

44. Ibid. (1949). p. 7.

45. "The Mwata Yamvo and his great feudatories are too inclined to regard the local administration as a negligible quantity. In its pride, the Lunda court views the head of a territoire merely as the lowest echelon in the administration while being convinced that [the Mwata Yamvo's] own essence emanates from a hierarchical apex." Territoire de Kapanga, Rapport AIMO (1948), p. 3. "The Mwata Yamvo has an inflated opinion of himself and thinks that he is too important a person to deal with the head of a mere territoire or his subordinates." Ibid. (1949), p. 7.

46. Territoire de Kapanga, Rapport AIMO (1947), p. 4; ibid. (1949), p. 3.

47. Territoire de Kapanga, Rapport AIMO (1947), p. 5; ibid. (1948), p. 3; ibid. (1949), p. 9.

48. Territoire de Kapanga, Rapport AIMO (1948), p. 9.

49. Included in the 26,000 francs collected in 1949 were 6,036 francs received by native circuit judges dispatched from the Musuumb. These fees were abolished in 1950. In addition to the thirteen new subordinate courts organized in 1950, two subordinate courts had been maintained at Kapanga and Tshibingu ever since these former sous-chefferies had been absorbed into the Mwaant Yaav's bailiwick in 1935. Their combined receipts also dropped—from 13,467 francs in 1949 to 6,117 francs in 1950. Territoire de Kapanga, Rapport AIMO (1950), p. 5.

50. Territoire de Kapanga, Rapport AIMO (1950), pp. 26-27.

51. Ibid., pp. 27-28.

52. Territoire de Kapanga, Rapport AIMO (1951), p. 23.

53. Based in part on "Le Mwata Yamvo Ditende . . . fait sa première discour [sic] a ses auditeurs" (n.d., 1951?) (ATK); "Oraison funèbre de Mwanta Yamvo Mbako Ditende Yawa Naweji III" by the Administrateur de territoire for Kapanga (2 June 1963); interviews with Mwaant Yaav Muteb II Mushid, Rev. William Davis, and Rev. Marvin Wolford (April 1972). See also L'Essor du Congo, 20 June 1951.

54. Procès-verbal de la séance du Conseil des Notables tenue à Musumba le 31 mai 1951 . . . (ATK).

55. "Note sur la succession du Grand Chef Mwanta Yamvo" (n.d., June 1963?), (ATK). Rev. Brinton, of the Methodist mission, was reportedly consulted by the Belgian authorities at the time of the succession and he strongly recommended Ditende even though the Tshombe family (themselves good Methodists) was backing Mushidi. (Interview with Rev. William Davis, April 1972.)

56. Territoire de Kapanga, Rapport AIMO (1951), p. 23.

57. Ibid. (1952), pp. 23-24; Territoire de Kapanga, Procès-Verbal Administratif du 6 décembre 1952 (ATK); interviews with Kanampumb Louis and with Rev. André Naweej (April 1972). Mushidi was subsequently permitted to live at Sandoa with his nephews.

58. Governor of Katanga to Director general of BCK, 21 May 1952.

7. From Tribalism to Ethnicity:
The Growth of Lunda and Cokwe National Sentiments

1. Territoire de Kapanga, Rapport AIMO (1952), p. 20; ibid. (1953) p. 29; ibid. (1955 to 1958), passim.

2. Young, *Politics in the Congo,* p. 29. Territoire de Kapanga, Rapport AIMO (1954), p. 25.

3. "Handsome Lord"—a pseudo-African praise name coined for the occasion by the colonial government's imaginative public relations and information agency.

4. As late as 1948, the Belgian administration was reportedly alarmed when, during his first trip to Belgium, Joseph Kapenda (father of Moïse Tshombe and of the two Lunda paramount chiefs elected in 1965 and 1973) arranged for the import of some old-fashioned flintlocks ("poupous") and gunpowder commonly used by African hunters. Mwaant Yaav Kaumba himself was suspected in connection with this episode. (Interview with Mwaant Yaav Muteb II Mushid, April 1972.)

5. INFORCONGO, *Le Congo Belge* (Brussels, 1958), II, pp. 20-21.

6. Ordonnances-Lois, 12 and 20 July 1945.

7. On the *paysannats,* see notably G. Malengreau, *Vers un Paysannat Indigène* (Brussels, Van Campenhout, 1949); M. Merlier, *Le Congo, de la colonisation belge à l'independance* (Paris, Maspero, 1962); "A propos des problèmes du milieu rural congolais," *Bull. du CEPSI,* March 1957. As early as 1949, Malengreau had noted that the *paysannats* were being regarded as "champs de l'Etat" (the term used by Africans to describe the acreage kept under government-imposed crops) and that the threat of being sent off to work on a *paysannat* was often used by foremen, teachers, etc. (p. 35, note 48).

8. The administrative headquarters for the territory was moved back from Malonga to Dilolo in 1949.

9. Territoire de Dilolo, Rapport AIMO (1951), p. 26.

10. Railroad companies were usually conceded sole mineral rights over a strip of land extending on either side of the rail line.

11. Territoire de Kapanga, Rapport AIMO (1955), p. 24.

12. Territoire de Kapanga, Rapport AIMO (1954), p. 18. See also Biebuyck, "Fondements," p. 806.

13. Territoire de Kapanga, Rapport AIMO (1955), p. 23. The son of a former *Rweej,* Kanampumb Louis was still holding this position in 1973.

14. Ibid. (1952), p. 24.

15. Territoire de Kapanga, Enquête politico-foncière sur le dégorgement de Musumba à base de paysannat indigène (1952), quoted in Biebuyck, "Fondements," pp. 816-817.

16. Territoire de Kapanga, Rapport AIMO (1953), p. 26; ibid. (1954), p. 22.

17. On the basis of his fieldwork in 1959-1960, Crine estimated the population of the Musuumb at 4,000. Current estimates range from 12,000 to

15,000. (Interviews with Mwaant Yaav Muteb II Mushid, Rev. William Davis, and Father Martin, April 1972.)

18. Interview with Rev. André Naweej, April 1972.

19. Territoire de Dilolo, Rapport AIMO (1956), p. 34.

20. Territoire de Kapanga, Rapport AIMO (1957), p. 21. She was subsequently forced to relinquish that position under pressure from the Zambian authorities.

21. For the role played by similar groups in attempting to modernize the Kongo kingship, see John Marcum *The Angolan Revolution,* vol. I (Cambridge, Mass., MIT Press, 1969), pp. 56-64.

22. This was later expressed by Cokwe leader Ambroise Muhunga in the form of a complaint that "our race is the only one that is [held as] inferior to other Congolese races" (Muhunga to Administrateur de territoire for Dilolo, 22 January 1957).

23. See Territoire de Sandoa, Rapport AIMO (1945), p. 20; ibid. (1949), p. 4; ibid. (1952), p. 34. "Chief Samutoma and his family . . . stress at every turn their separatism vis à vis the Mwata Yamvo. . . . He represents himself as the champion and the representative of the Cokwe nation."

24. Chief Tshisenge of Dilolo and his neighbor, Chief Kandala, were at loggerheads over several issues; the first had no clear traditional base and the second led a minority clan. As for Sakundundu (who had fewer subjects than the other three), he was described in one report as suffering from "precocious cretinism caused by excessive drinking." Territoire de Sandoa, Rapport AIMO (1958), p. 31.

25. Over the three-year period 1954-1956, the excess of immigration over the emigration for Dilolo territory resulted in a population gain of 3,493. Of this gain, 36% was registered by the two Cokwe chefferies of Tshisenge and Kandala alone; this represents more than five times the population increase due to excess immigration over the same three years for the entire chefferie of the Mwaant Yaav. Compiled from statistics in Territoire de Dilolo, Rapport AIMO (1956), p. 21, and Territoire de Kapanga, Rapport AIMO (1958), p. 13.

26. Territoire de Dilolo, Rapport AIMO (1957), pp. 31-34.

27. Ibid. (1955), p. 35; ibid. (1956), p. 32.

28. Administrateur de territoire de Sandoa, "Note sur l'origine, l'existence, l'activité et le but des associations indigènes en Territoire de Sandoa" (1958) (ATS).

29. District commissioner for Lualaba to Vice-governor general for Katanga, 29 January 1960. Ambroise Muhunga, leader and founder of ATCAR, the Cokwe political party, offered (as might be expected) much larger estimates: 75% in Sandoa territory and 95% in Dilolo territory. ("La grande nation des Tshokwe et sa dynastie," MRAC). The 1960 breakdown of Cokwe population by chefferie in Sandoa territory was reported to be: Samutoma and Sakundundu, 100%; Lumanga, 65%; Tshipao, 65%; Mbako, 55%; Kayembe Mukulu, 50%; Muteba, 45%; Tshibamba, 35%. District commissioner for Lualaba to Governor of Katanga, 8 March 1960.

30. District commissioner's authorization of 19 October 1955.

31. District commissioner's authorizations of 8 May 1956 (Jadotville), 2 March 1957 (Shinkolobwe), and 24 July 1957 (Lubudi).

32. Administrateur de territoire for Sandoa, "Note sur l'origine, l'existence, l'activité et le but des associations indigènes en Territoire de Sandoa" (1958?). Province du Katanga, Service des AIMO, "Note synthétique concernant l'Association des Tshokwe" (4 June 1958).

33. Muhunga to Governor of Katanga. 22 January 1957.

34. Marcum, *The Angolan Revolution,* p. 117.

35. Administrateur de territoire for Dilolo to Muhunga, 16 January 1957.

36. Marcum, *The Angolan Revolution,* pp. 117-118. A certain overlap seems to have existed between the two associations. In December 1957 the chairman of the Dilolo branch of Atcar was found to be carrying a membership card of the Ukwashi wa Chokwe signed by John Kazila (sic). District commissioner for Lualaba to Governor of Katanga, 12 December 1957.

37. Territoire de Dilolo, Rapport AIMO (1957), p. 29.

38. Territoire de Sandoa, Rapport AIMO (1957), pp. 31-32.

39. Territoire de Dilolo, Rapport AIMO (1957), pp. 28-29.

40. Territoire de Kapanga, Rapport AIMO (1957); interviews with Mwaant Yaav Muteb II Mushid and Rev. William Davis (April 1972). Ditende was reportedly fearful of being poisoned or otherwise harmed by his enemies at that time (and indeed until his death), so strong was the animosity between the two factions.

41. "The Mwata Yamvo," wrote the administrateur for Sandoa, "is personally hostile to any kind of association." ("Note sur l'origine.")

42. Territoire de Dilolo, Rapport AIMO (1957), p. 28; Territoire de Sandoa, Rapport AIMO (1957), p. 29. Province du Katanga, Service des AIMO, note to Governor of Katanga, 27 March 1958. Idem, note to Provincial commissioner, 23 April 1958. Provincial commissioner (pro Governor of Katanga) to Governor general, 28 June 1958.

43. Muhunga to Governor of Katanga, 10 May 1958.

44. Governor of Katanga to District commissioner for Lualaba, 25 November 1957.

45. Ordinance of 11 February 1926.

46. See, e.g., Territoire de Dilolo, Rapport AIMO (1957), pp. 29-30; Territoire de Sandoa, Rapport AIMO (1957), p. 29.

47. The following year, the procedure was extended to the four provincial capitals of Coquilhatville (Mbandaka), Stanleyville (Kisangani), Luluabourg (Kananga), and Bukavu.

48. It was estimated in 1956 that 26.8% of the adult male African population of Elisabethville consisted of Kasai BaLuba and another 18.1% of Shankadi (Katanga) BaLuba. J. Denis, "Elisabethville: Matériaux pour une étude de la population africaine," *Bull. du CEPSI,* no. 34 (1956), p. 167. In 1958, 52% of the African manpower employed in Elisabethville originated from outside Katanga (including 18.1% from outside the Congo itself). Ville d'Elisabethville, Service de la Population Noire, Rapport 1958.

49. Denis, "Elisabethville," p. 167, and "Jadotville: Matériaux pour une étude de la population africaine," *Bull. du CEPSI,* no. 35 (1956), p. 45. Percentages are of the total adult male African population, including foreign Africans. Figures for Jadotville cover only the area subsequently incorporated under the name Kikula "Commune," the only one where elections were held in 1957.

50. The rest of the African population either lived in one of two company camps (Union Minière and BCK) or consisted of domestic servants housed by their employers in the European section of town.

51. One example of this sort of fragmentation was Ward 6, where a candidate from the small Zela tribe won the seat with only 22.8% of the vote, while the four Kasai BaLuba candidates split 50.1% of the vote.

52. Territoire de Dilolo, Rapport AIMO (1957), p. 29. "This is indispensable," commented the District commissioner, "and should be the major preoccupation of the administrateur during 1958 with respect to this issue. As a matter of fact, it would be desirable that this ringleader [Muhunga] not appear in the Tshokwe areas any more." Observations du Commissaire de district on the above.

53. Van M. . . . to Governor of Katanga, 28 March 1958 and 1 May 1958; Van M. . . . to Provincial commissioner, 7 May 1958; Van M. . . . to Director general of BCK, 3 July 1958; Van M. . . . to Mwaant Yaav, 28 March 1958. Territoire de Dilolo, Rapport AIMO (1958), pp. 27, 30. Later incidents involving Chief Tshisangama, Muhunga, and others are better documented.

54. Territoire de Dilolo, Rapport AIMO (1958), pp. 28-29.

55. District du Haut-Lomami, Procès-verbal de la réunion-palabre des chef coutumiers du Territoire de Sandoa tenue le 28 mars 1958 sous la présidence du Commissaire de District.

56. The administration tried to minimize the psychological effects of Samutoma's claim by affecting to treat the title of Mwatshisenge as a reigning name (similar to Ditende's assumption of the name of Yaav Naweej), but Samutoma insisted that the name of Mwatshisenge be officially applied to his chefferie, as it was when it had first been organized in the 1920s. Administrateur de territoire for Sandoa to District commissioner for Lualaba, 9 July 1959.

57. "From a political standpoint, we consider the attempts to create a Tshokwe association at the provincial level, or even covering a whole territoire, as dangerous." Provincial commissioner pro Governor of Katanga to Governor general, 28 June 1958.

58. Provincial commissioner to Muhunga, 5 May 1958. Muhunga to Governor of Katanga, 10 May 1958. Governor of Katanga to Muhunga, 6 June 1958. Service provincial des AIMO (Katanga), "Note synthétique concernant l'Association des Tshokwe," 4 June 1958.

59. Provincial commissioner (pro Governor of Katanga) to Ambroise Muhunga, 6 June 1958; Service Provincial des AIMO, "Note synthétique concernant l'Association des Tshokwe" (4 June 1958).

60. This declaration, together with a message from King Baudouin, was

eventually delivered on January 13, 1959 — but not before the Congo had been shaken to its foundations by the Léopoldville riots of January 4-6, 1959.

61. Munongo was a civil servant at the time, and as such was disqualified from leading a political organization.

62. *L'Essor du Congo*, 26 May 1959. This had been for years the position of settler groups.

63. *Congo 1959* (Brussels, CRISP, 1960), pp. 117-118.

64. Fédération des Associations de Ressortissants de la Province du Kasai, led by Isaac Kalonji. It was founded on December 1, 1958, to fill the gap left after the Fegebaceka had been dissolved in November — partly to appease the "authentic Katangese."

65. Territoire de Dilolo, Rapport AIMO (1958), p. 27.

66. This does not include mining, which was not an important source of employment in Elisabethville (4.35% of jobs in 1956).

67. Ville d'Elisabethville, Service de la Population Noire, Rapport 1958, passim. Another facet of the economic crisis, which helps explain some of the animosity felt by "authentic Katangese" is reflected in the fact that 62% of the jobs lost in Elisabethville from 1956 to 1958 were lost by workers originating from the rural areas of Katanga, although that group accounted for less than half of the total African manpower.

68. In 1956 fewer than 22% of the adult men from Lunda residing in Jadotville originated from the Kapanga area. Compiled from Denis, "Jadotville," p. 44.

69. Territoire de Sandoa, Rapport AIMO (1958), p. 28.

70. A similar phenomenon was noted in the Kwilu by Herbert Weiss in his *Political Protest in the Congo* (Princeton, N. J., Princeton University Press, 1968) and in the Lower Congo by Vice-Governor Schöeller. See his "Rapport sur le situation dans le District des Cataractes" in *Congo 1959*, pp. 100-107.

71. Muhunga to District commissioner for Lualaba, 8 August 1959. The incident itself occurred in May.

72. Mwaant Yaav Ditend Yav a Nawej to Minister for Belgian Congo and Ruanda Urundi, 30 January 1959, reprinted in part in *Congo 1959*, p. 51.

73. For some of the views expressed by these circles on the role of traditional chiefs, see the documents published in *Congo 1959*, pp. 22-28, 31-34, 37-39.

74. District commissioner for Lualaba to Administrateurs de territoire for Sandoa and Kapanga, 3 July 1959. Idem to Governor of Katanga, 6 August 1959. For the address delivered by the Mwaant Yaav on this occasion, see *Congo 1959*, p. 53.

75. Administrateur de territoire for Sandoa to District commissioner for Lualaba, 9 July 1959. The request was denied by the District commissioner in his reply of 22 July "in the interest of maintaining peace."

76. District commissioner for Lualaba to Governor of Katanga, 6 August 1959.

77. District commissioner for Lualaba to Governor of Katanga, 20 January 1960. European settlers (UCOL) and veterans' groups (UFAC) also

showed their preferences by organizing an elaborate reception for the Mwaant Yaav in Kolwezi and by presenting him with a set of expensive hunting rifles.

78. Muhunga to District commissioner for Lualaba, 8 August 1959.

79. Acting Governor for Katanga to Governor general, 9 October 1959.

80. Eventually, the Kasai Luba community in Katanga wound up being divided into three major factions: the Fedeka, which stayed in the Cartel; the MNC-Kalonji, which won one of six seats from Elisabethville to the Katanga Provincial Assembly in 1960; and the Fédération Générale du Congo, an offshoot of the former Fegebaceka, led by André Kadima — not to mention a handful of individuals like Joseph Mbuyi (member of Lumumba's first cabinet) who remained faithful to the MNC-Lumumba. Similar ethnic considerations drove the much smaller Lulua immigrant community of Katanga into the arms of Conakat.

81. For Lumumba's version of these incidents, which he attributed to Belgian provocations, see *Congo 1959,* pp. 198-199.

82. District commissioner for Lualaba to Administrateurs de territoire for Kolwezi, Sandoa, and Dilolo, 8 December 1959, App. (original in KiCokwe).

83. Ibid.

84. This is not to suggest that the rural populations were actually docile — quite the contrary as it turned out, at least in some areas — but the urban elites as well as the administration believed, each in their own way, that the chiefs and other rural notables could and would deliver the "bush vote" to the so-called moderates.

85. Territoire de Dilolo, Rapport AIMO (1959), p. 3. District commissioner for Lualaba to Governor of Katanga, 29 January 1960.

86. This was in accordance with the administration's efforts to keep party politics as much as possible out of the local government elections. This did not deter Tshombe from claiming that Conakat had won 426 out of 484 seats in Katanga by simply adding to the 84 seats it had carried in its own right the 342 seats won by traditional, local, and independent candidates. *Congo 1960,* p. 228.

87. District commissioner for Lualaba to Governor for Katanga, 20 January, 1960.

88. Rapport Sûreté sur la réunion de l'Atcar tenue au Tribunal du C.E.C. de Kolwezi le 10 janvier 1960 (13 January 1960).

89. Mwaant Yaav . . . to Minister for Belgian Congo and Ruanda-Urundi, 31 January 1959.

90. Note from Chef du Service provincial des AIMO to Governor for Katanga, 2 February 1959.

91. Administrateur de territoire for Kapanga to Governor of Katanga, 26 February 1959.

92. Provincial commissioner (pro Governor of Kasai) to Governor of Katanga, 22 April 1959.

93. The Abako had been dissolved after the January riots and its leaders, Kasa-Vubu, Kanza, and Nzeza, imprisoned, then flown to Brussels (March

14) where they remained until mid-May. In their absence, a clandestine *Mouvement de Résistance Bakongo* assumed a clearly separatist position. See, e.g., their declaration of April 23, 1959, in *ABAKO 1950-1960* (Brussels, CRISP, 1963), pp. 203-205.

94. Governor of Léopoldville Province to Governor general, 1 April 1959.

95. Governor general to Governor of Katanga, 26 May 1959.

96. Acting Administrateur de territoire for Kapanga to Governor of Katanga, 2 July 1959, and comments by District commissioner for Lualaba, 8 July 1959. Provincial commissioner for Katanga to Governor general, 29 July 1959.

97. Governor general to Governor of Katanga, 19 November 1959. Governor of Katanga to District commissioner for Lualaba, 1 December 1959. It was the Kiamfu, however, who eventually visited the Mwaant Yaav in July 1961, during the Katanga secession.

98. Mwaant Yaav Ditend Yav a Nawej III to Governors of Léopoldville and Kasai Provinces, Governor of Saudimbo (= Va. Henrique de Carvalho?), District Commissioner of Saleze (= Solwezi), Administrators [sic] for Balovale, Mwinilunga, and Kawamba, Administrateurs de territoire for Dilolo and Kolwezi, 8 January 1960. Governor of Katanga to Governors of Léopoldville and Kasai, 4 February 1960; idem to British Vice-consul at Elisabethville, 5 February 1960. One reply mentions their number in Northern Rhodesia as 74,328, plus 12,351 Luchazi, Cokwe, Lamba, Mbundu, Kaonde, and Baluba who also reportedly acknowledged the Mwaant Yaav's traditional paramountcy. (British Vice-consul at Elisabethville to Governor of Katanga, 11 May 1960.)

99. Territoire de Dilolo, Rapport AIMO (1959), p. 4. District commissioner for Lualaba, 20 January 1960, 29 January 1960, 8 March 1960.

100. Administrateur de territoire for Sandoa to District commissioner for Lualaba, 19 January 1960.

101. Interviews with Mwaant Yaav Muteb II Mushid, Nswaan Mulopw Mutshaila Ditende, Kanampumb Louis, Rev. André Naweej, Rev. William Davis (April 1972). Samuel Mawawa, Maurice Kangaji, and Gabriel Kanyimbu Kambol have been members of the national legislature.

102. District commissioner for Lualaba to Governor of Katanga, 20 January 1960, 29 January 1960.

103. Biebuyck, "Fondements," pp. 816-817. These views were sharply criticized in the editorial introduction to Leon Duysters' "Histoire des Aluunda" in *Problèmes d'Afrique Centrale* (1958), II, pp. 75-98, the journal published by the alumni of the Institut Universitaire des Territoires d'Outre-Mer (INUTOM), which reflected the views of the colonial service.

104. The only exception, of course, being the indirect use of the term in the full name of Gassomel. In the late fifties, Moïse Tshombe had also sponsored an amateur soccer team in Elisabethville under the name of "Empire Lunda."

105. Lumanga, Muteba, Tshisangama, Sakayongo, Tshisenge, Kandala, Samutoma, and Katende.

106. District commissioner for Lualaba to Governor of Katanga, 25 February 1960.

107. *Pourquoi Pas? Congo*, 6 (February 1960), pp. 79-81. Muhunga had expressed similar views before in his unpublished manuscript, "La grande nation des Tshokwe," where he wrote, "In the Tshokwe language the word *Lunda* means: a slave who may be struck if he does not work well" (p. 36).

108. Comité central du Gassomel to Procureur général at Elisabethville, 11 February 1960.

109. Mwata Yamvo Yav a Nawej III to Governor general, 25 February 1960. See Appendix 1.

110. Territoire de Kapanga, Procès-verbal de la réunion du 2 mars 1960.

111. Assistant District commissioner for Lualaba, Rapport de route (5 March 1960). District commissioner for Lualaba to Governor of Katanga, 8 March 1960.

112. That latter conflict was more in evidence later during the Kamina incidents of May 22, 1960. It involved attacks against the immigrant BaLuba by the subjects of Kasongo Nyembo who, unlike most other Luba chiefs in Katanga, had rallied to Conakat, not to the Cartel.

8. Decolonization, Secession, Reunification: The Lunda and Cokwe through the Congo Crisis

1. See the analysis of voting patterns at the Round Table in Georges Henri Dumont, *La Table Ronde Belgo-Congolaise* (Brussels, Editions Universitaires, 1961), passim.

2. Célestin Sapindji succeeded Samutoma on 29 May 1968, under the title of Mwene Mwatshisenge Lukasa I. (Ndua, "L'Installation des Tutshokwe," p. 32.) He has also been a member of the national legislature.

3. Ernest Koji, Oscar Mbundj, Daniel Tshombe and Amédée Chisol.

4. Because of the large number of BaLuba from Kasai and Katanga working in Kolwezi, Atcar ran as a member of the Cartel, and not on its own as it did in Dilolo Territory. All percentages are adapted from *L'Etoile-Nyota* (Elisabethville), 26 May 1960.

5. Such distortions were common throughout the Congo and resulted from the fact that proportional representation was being applied within relatively small multi-member constituencies (two to six members in Katanga) thus limiting its mathematical accuracy. The House seat won by Muhunga was carried by 29,657 votes, while Conakat seats throughout Katanga were won with an average vote of 13,109.

6. Following the incident of the *Pourquoi Pas? Congo* interview (see Chapter 7), the decision restricting Muhunga's travels in the southwest had of course been reiterated and was lifted only at the end of March after an intervention from the Collège Exécutif Provincial. (Muhunga to District commissioner for Lualaba, 3 March 1960. Governor of Katanga to District commissioner for Lualaba, 24 March 1960.)

7. Another cause for the Cartel's disgruntlement was the high number of

irregularities allegedly committed during the election itself. Of all the official protests lodged by the Cartel against such practices, none was given redress by the administration. For this and related developments, see J. Gérard-Libois, *Katanga Secession* (Madison, Wis., University of Wisconsin Press, 1966), pp. 63-89.

8. As representative of the central government, the State commissioner was empowered to override provincial authorities over certain issues and to apply a suspensive veto to some of their decisions. The importance attached to this post by the Cartel is attested by the fact that it had also been offered to Sendwe five days earlier when Kasa-Vubu had attempted unsuccessfully to organize an alternative coalition. *Congo 1960*, p. 254.

9. Tshombe, Munongo, Kibwe, Kiwele, and others had all run for the Provincial Assembly, not for the House of Representatives.

10. Not to be confused with the Province of Lualaba created out of the former District of Lualaba on June 30, 1963, after the end of the secession.

11. In its original proclamation, the "Province of Lualaba" was described as including the districts of Upper Lomami, Tanganyika, *and* Lualaba (i.e., the Territories of Kapanga, Sandoa, Dilolo, Kolwezi, and Lubudi) but, needless to say, the authority of the Manono government never extended over the slightest portion of the latter district.

12. Of late, Muhunga has been leading an obscure faith-healing church based in Kinshasa, the "Communauté Evangélique Guérison par le Christ."

13. Radio-Léopoldville broadcast, 14 August 1960, quoted in *Congo 1960*, p. 733.

14. According to a Luba informant, Cokwe refugees were sometimes viewed with suspicion by other residents of the "Baluba Camp," not only because they did not speak Luba, but also because their ethnic affinities with the Lunda made it possible for the Katanga authorities to infiltrate their ranks with "Cokwe-ized Lunda" indicators.

15. Article 87.

16. *L'Essor du Congo*, 18 July 1960.

17. Chief Mwenda Munongo of the BaYeke (brother of Interior Minister Godefroid Munongo), Paramount Chief Kasongo Nyembo (Emmanuel Ndaie), Paramount Chief Manda Kaseke of the BaTabwa, and Chief Mbako.

18. These were the Mwaant Yaav, Tshisenge, Kabongo, Kasongo Nyembo, Mwenda Munongo, Katanga Kianana, Mpande (of the BaSanga), Tumbwe (of Albertville), Lengwe Manteka (of Nyunzu), and Marco Kilanga. *L'Essor du Congo*, 18 July 1960.

19. Constitution de l'Etat du Katanga (*Moniteur Katangais*, 8 August 1960), Articles 31 and 32.

20. Etat du Katanga, Ministère de l'intérieur, Procès-verbal de la Réunion inter-bureaux du 17 janvier 1961.

21. Ibid., 12 January and 18 January 1961.

22. In January 1961 the effectives of the *service territorial* for Katanga numbered 28 Africans and 22 expatriates. Ibid., 27 January 1961.

23. Quoted in Gérard-Libois, *Katanga Secession*, p. 205.

24. Mwanta Yanvo to Minister of the interior (State of Katanga), 14 November 1962. The request was not answered until after the end of the secession, at which time Munongo (now acting as Minister of interior of the Province of South Katanga) indicated that the emoluments of a chief were "legally based on the importance of the population of his chefferie." He went on to suggest with veiled irony that the Mwaant Yaav's "authority and spiritual influence" might be concretized by voluntary tribute from his far-flung subjects, but that the government could not intervene to force the recovery of such contributions (Minister of interior of South Katanga to Grand Chef Mwanta-Yanvo, 6 February 1963.

25. Note de M. Gomez sur la situation à Musumba (Kolwezi, 30 September 1960).

26. Administrateur de territoire for Kapanga to District commissioner for Lualaba, 4 November 1960. The Mwaant Yaav eventually did make an extensive trip through the United States.

27. In June 1961 Delvaux also played a significant role in securing the release of Tshombe, who had been arrested at the Coquilhatville conference.

28. Note sur le séjour a Kapanga de M. Gomez (Elisabethville, 13 October 1960).

29. The strong hostility felt by the Katanga authorities toward Kenneth Kaunda's UNIP (and the support they gave to Harry Nkumbula's ANC) were enough in themselves to ensure that such facilities would no longer be available when Zambia became independent. In one document, UNIP is described as "extremist, Lumumbist, and even Communist" (Procès-verbal de la Réunion inter-bureaux, Ministère de l'intérieur, 7 January 1961).

30. In July 1964, following Tshombe's appointment as prime minister of the Congo, Ditende's successor, Gaston Mushidi, made a first trip to Angola (Teixeira de Sousa) with Thomas Kabwita Tshombe, apparently to arrange for the return to the Congo of the Katanga regiment led by Colonel Tshipola, which mutinied at Kisangani two years later (Administrateur de territoire for Dilolo District commissioner for Lualaba, 12 September 1964). I found no record of any official trip to Angola by Ditende himself.

31. See Appendix 1

32. Note sur le séjour a Kapanga de M. Gomez.

33. Paul Katanga, *Le Progrès* (Léopoldville), 13 November 1962, quoted in J. C. Willame, *Les Provinces du Congo*, vol. 2 (Léopoldville, Université Lovanium, 1964), p. 44.

34. Article 7 of the Loi Fondamentale was amended to that effect on March 9, 1962.

35. See *Congo 1963*, pp. 377-384.

36. House, 21 May 1963; Senate, 28 May 1963.

37. Among the candidates who were being considered, at least nominally, were the *Nswaan Mulopw* Mutshaila Ditende, Chief Dumba and Muteba, son of Kaumba (interviews with Mwaant Yaav Muteb II Mushid, Nswaan Mulopw Mutshaila Ditende and Rev. André Naweej, April 1972).

38. Territoire de Kapanga, Note sur la succession du Grand Chef Mwanta

Yambo (n.d., 1963?). The "other member of the same line" mentioned by Mbundj was Moïse Tshombe's brother David Yav, who subsequently became Mwaant Yaav Muteb II Mushid (1965-1973).

39. Conakat member of the House of Representatives in 1960, then a member of the Katanga National Assembly. Later held the position of Minister of the Interior of the national government.

40. After a stormy period in power, Jason Sendwe was eventually overthrown and assassinated as the so-called Mulelist rebellion spread into north Katanga (19 June 1964).

41. During that ceremony, he adopted the official name of Lumanga Kawele. Mushidi's ritual investiture was followed by the formal installation of Thomas Kabwita as Chief Lumanga.

42. As abstracted in Territoire de Kapanga, Rapport sur l'investiture du Grand Chef Mwant Yamvo Mushidi (18 November 1963).

43. Administrateur de territoire for Kapanga to Direction du service territorial of Lualaba Province, 19 November 1963.

44. Administrateur de territoire for Kapanga to Secretary of the Mwant Yamvo, 29 November 1963.

45. A summary of Tshombe's strategy will be found in *Congo 1964*, pp. 123-142, 148-160.

46. In a press communiqué released on June 27, Tshombe stated matter-of-factly, "I did not return too early. I did not return too late. I returned opportunely." Quoted by Pierre Davister in *Pourquoi Pas?*, 3 July 1964.

47. Anonymous report on "the situation in Kolwezi," June 1964, quoted in *Congo 1964*, p. 153. It would appear that this document was planted by pro-Tshombe elements to intimidate the Léopoldville authorities.

48. Administrateur de territoire for Dilolo to District commissioner for Lualaba, 12 September 1964.

49. *Le Courrier d'Afrique*, 21 July 1964.

50. Administrateur de territoire for Dilolo to District commissioner for Lualaba, 12 September 1964.

51. Commissaire de police for Kisenge, Report (3 October 1964).

52. The initiative to organize Conaco had indeed stemmed from the Léopoldville section of Conakat—a sort of "branch office" of a party whose only real base was in Katanga. In Lualaba and Katanga Oriental, local party leaders simply ignored the name Conaco and ran under the Conakat label.

53. Later Minister of external trade, then of social affairs under the Mobutu regime.

54. For Lualaba, 21,611 votes; for S. Kasai, 3,637; blank or void, 1,339.

55. Ordinance of 6 July 1965 (*Moniteur Congolais*, no. 13, 1 July 1965).

56. *Le Courrier d'Afrique*, 19 July 1965.

57. A few weeks earlier, Thomas Kabwita had been awarded an advance consolation prize in the form of a seat on the board of Union Minière. Three other Congolese were also appointed at that time, one of them being Munongo's brother, Chief Mwenda Munongo.

58. Mwasaza later served as secretary to the late Mwaant Yaav, Muteb II Mushid.

59. Commandant de police, Secteur Lualaba to Minister of the interior for Lualaba, 13 August 1965.

60. Also considered for the succession was Ditende's son, Chief Mbako.

61. Ian Colvin, *The Rise and Fall of Moïse Tshombe* (London, Frewin, 1968), pp. 228, 230.

62. Upon investigation, that particular incident turned out to involve a harmless enough visit by four Belgian officers of the Miba diamond-mining company from Kasai. Rapport de la Commission parlementaire d'enquête au Sud-Katanga, quoted in *Congo 1966,* p. 368.

63. This account of the April 1967 incidents is based on interviews with Rev. Marvin Wolford and Mwaant Yaav Muteb II Mushid (April 1972).

64. Like all other administrateurs de territoires, the Mwaant Yaav was subsequently put in charge of the local branch of the Mouvement Populaire de la Révolution (MPR), the country's single political party.

9. Conclusion

1. A similar view is held by Ian Cunnison regarding the Kazembe kingdom. See his "Kazembe and the Portuguese, 1789-1832," *Journal of African History,* 2 (1961), p. 65.

2. The case of the Igbo is analyzed by A. C. and D. R. Smock in their "Ethnicity and Attitudes Toward Development in Eastern Nigeria," *Journal of Developing Areas,* July 1969, pp. 499-512.

3. An early version of this phenomenon is found in the granting of feudal titles to the dignitaries of the Kingdom of Kongo by the king of Portugal.

4. Robert Delavignette, *Freedom and Authority in French West Africa* (London, Oxford University Press, 1950), p. 82 ("Do not let us be afraid to sweep Africa clean of its feudalists by officializing them," etc.). A similar attitude was expressed, in a wholly different context, by British officials in Northern Nigeria; e.g. Sir Donald Cameron, *My Tanganyika Service and Some Nigeria* (London, Allen & Unwin, 1939). These views are still found in recent assessments of the colonial era; e.g. Robert Heussler, *The British in Northern Nigeria* (London, Oxford University Press, 1968) pp. 23, 76, and passim.

5. Jacques Lombard, *Structures de type "féodal" en Afrique noire* (Paris, Mouton, 1965), pp. 371-372.

6. Jacques J. Maquet, "Une hypothèse pour l'étude des féodalités africaines," *Cahiers d'Etudes Africaines,* II, no. 6 (1962), pp. 292-314.

7. Aidan Southall, *Alur Society: A Study in the Processes and Types of Domination* (Cambridge, Heffer, 1956), pp. 250-251.

8. Dias de Carvalho, *Ethnographia e Historia,* p. 62.

9. Biebuyck, "Fondements," pp. 791-797.

10. In recent years, the cases of Tshipao and Mbako show that this type of situation did not disappear with the advent of colonial rule.

11. Jan Vansina, Raymond Mauny, and L. V. Thomas, *The Historian in Tropical Africa* (London, Oxford University Press, 1964), p. 85. In the words of Conrad P. Kottak, "mediation of exchange of products from different eco-

logical zones has been the foundation of most African states" ("Ecological Variables in the Origin and Evolution of African States: The Buganda Example," *Comparative Studies in Society and History*, 1972, p. 377).

12. Royal monopoly of certain prestige goods (e.g., leopard skins) was clearly not of an economic nature.

13. Robert F. Stevenson, *Population and Political Systems in Tropical Africa* (New York, Columbia University Press, 1968), p. 232 and esp. Chapter V, which deals with the Bemba.

14. V. Vermeulen, *Deficiences et dangers de notre politique indigène* (Brussels, Impr. Veuve Monnom, 1952); Biebuyck, "Fondements," pp. 816-817; Georges Brausch, *Belgian Administration in the Congo*, pp. 46-47.

15. See Hubert Deschamps, "Et maintenant, Lord Lugard?" *Africa*, XXXIII (1963), pp. 293-305.

16. D. A. Low and R. C. Pratt, *Buganda and British Overrule, 1900-1955* (London, Oxford University Press, 1960), p. 301.

17. See Robert Heussler, *The British in Northern Nigeria*, pp. 125-129, 185, and passim.

18. Ralph A. Austen, *Northwest Tanzania Under German and British Rule* (New Haven, Conn., Yale University Press, 1968), p. 254.

19. David E. Apter, *The Political Kingdom in Uganda* (Princeton, N. J., Princeton University Press, 1961), pp. 85-89.

20. J. Gus Liebenow, *Colonial Rule and Political Development in Tanzania: The Case of the Makonde* (Evanston, Ill., Northwestern University Press, 1971), p. 334.

21. Aristide Zolberg, *Creating Political Order: The Party States of West Africa* (Chicago, Rand McNally, 1966), p. 141. The concept of patrimonialism has been convincingly applied to the study of Congolese politics in a recent study by Jean-Claude Willame, *Patrimonialsim and Political Change in the Congo* (Stanford, Calif., Stanford University Press, 1972).

22. Willame, *Patrimonialism*, p. 34. See also p. 55.

23. Heussler, *The British in Northern Nigeria*, p. 83.

24. Zolberg, *Creating Political Order*, p. 144.

Bibliography

Official documents

General material
Bulletin Officiel de l'Etat Indépendant du Congo.
Bulletin Officiel du Congo Belge.
Bulletin Administratif du Congo Belge.
Moniteur Congolais.
Moniteur Katangais.
Codes et Lois du Congo Belge (compiled by P. Piron and J. Devos), 4 vols. (Brussels, F. Larcier, 1959).
Recueil à l'Usage des Fonctionnaires de l'Administration en Service Territorial (RUFAST) (Brussels, Ministère des Colonies, 1925) 4th. ed.
Gouvernement Général du Congo Belge. Section de la Documentation. *Documents pour servir à l'étude des populations du Congo Belge* (Léopoldville, 1958) Collection "Archives du Congo Belge," no. 2.

Periodical reports (Katanga)
District du Lualaba-Kasai. Rapports politiques (irreg.), AMC.
District du Kasai (Vice-Gouvernement Général du Katanga). Rapports annuels (1912-1913), AMC.
District de la Lulua (Vice-Gouvernement Général du Katanga). Rapports annuels (1913-1921, 1929-1932), APK. (Superseded by quarterly and semiannual reports, 1921-1928.)
District de la Lulua (Vice-Gouvernement Général du Katanga). Rapports trimestriels sur l'administration générale (also known as Rapports politiques and Rapports modèle B) (1913-1922), APK.
District de la Lulua (Vice-Gouvernement Général du Katanga). Rapports semestriels sur l'administration générale (also known as Rapports politiques and Rapports modèle B) (1922-1929), APK.
District du Lualaba (Province du Katanga). Rapports AIMO. Notes synthé-

tiques du commissaire de district sur la situation politique (1933-1945), APK.

Province du Katanga (also Province d'Elisabethville). Rapports AIMO (1934-1958), APK.

Territoire de Dilolo (Province du Katanga). Rapports AIMO (1945-1958), APK/ATD ("Territoire de Malonga," 1945-1948).

———. Rapports de la Sûreté (1945-1958), APK/ATD.

Territoire de Kapanga (Province du Katanga). Rapports AIMO (1947-1958), APK/ATK.

———. Rapports de la Sûreté (1947-1958), APK/ATK.

Territoire de Sandoa (Province du Katanga). Rapports AIMO (1945-1958), APK/ATS (covers Kapanga, 1945-1946).

———. Rapports de la Sûreté (1945-1958), APK/ATS.

Ville d'Elisabethville (Province du Katanga). Service de la population Noire. Rapports annuels (1956-1958), APK.

Bourse du Travail du Katanga/Office Central du Travail au Katanga. Rapports (irreg.), APK.

Rapport au Roi de la Sous-Commission du Katanga pour la Protection des Indigènes. *Bulletin Officiel* (1925), no. 4.

Other official documents

Binamé, A. *La consultation du 22 décembre 1957 dans la Commune de Kikula* (Ville de Jadotville, 30 April 1958), APK.

Duysters, Léon. *Ordre de marche et de combat de l'armée ruund* (Kapanga, 31 December 1925), ATK.

———. *Rapport de sortie de charge, Territoire de Kapanga* (Kapanga, 31 July 1927) followed by *Avis et considérations du commissaire de district,* AMC.

Enquête politico-foncière sur le dégorgement de Musumba à base de paysannat indigène (Kapanga, 1952), ATK.

Etat Indépendant du Congo. Courrier du Departement de l'Interieur, AMC.

Gomez, R. *Note sur la situation à Musumba* (30 September 1960), APK.

———. *Note sur le séjour à Kapanga de M. Gomez, R., conseiller du Mwata Yamfu* (13 October 1960), APK.

Gosme. *Notice sur les populations du District de la Lulua* (29 June 1913), APK/ATK.

Hupperts. *Historique de la chefferie du Mwata Yamvo* (Kapanga, 4 April 1917), APK/ATK.

Jammaers. *Rapport sur la délégation Lunda envoyée au Kwango en mission politique* (Kapanga, 9 April 1931), APK.

Katanga (Etat du). Ministère de l'Intérieur. *Procès-verbaux des réunions inter-bureaux* (Elisabethville, 1961-1962), APK.

Katanga (Province du). Service des AIMO. *Note du Chef du Service Provincial des AIMO sur l'Association des Tshokwe* (Elisabethville, 27 March 1958), APK.

―――. *Note du Chef du Service Provincial des AIMO sur les relations Lunda-Tshokwe* (Elisabethville, 23 April 1958), APK.

―――. *Note synthétique concernant l'Association des Tshokwe* (Elisabethville, 4 June 1958), APK.

Katanga (Vice-Gouvernement Général du). District de la Lulua. *Convention entre le Mwata Yamvo Kaumba et le Commissaire de District Van den Byvang* (Kapanga, 29 September 1923), APK.

Katanga (Province du). District du Haut-Lomami. *Rapport sur la réunion-palabre des chefs de Circonscriptions Indigènes du Territoire de Sandoa sous la direction du Commissaire de District du Haut-Lomami* (23 March 1958), APK.

Katanga (Province du). District du Lualaba. *Rapport sur la réunion des chefs Lunda et Tshokwe tenue à Kisenge le 14 février 1960* (25 February 1960), APK.

Katanga (Province du). Territoire de Kapanga. *Procès-verbal de la réunion du 2 mars 1960 . . . tenue à Kapanga* (Kapanga, 2 March 1960), APK.

Katanga (Vice-Gouvernement du). Territoire de Luashi. *Rapport d'enquête, Chefferie de Sakayongo* (1 September 1926), MRAC.

[Katanga] Province d'Elisabethville. Territoire de Malonga. *Chefferie de Tshisenge. Rapport d'enquête* (1935), MRAC.

―――. *Chefferie de Saluseke. Rapport d'enquête* (n.d.), MRAC.

―――. *Chefferie de Sakambundji. Rapport d'enquête* (n.d.), MRAC.

Katanga (Province du). Territoire de Sandoa. *Note sur l'origine, l'existence, l'activité et le but des associations indigènes en Territoire de Sandoa* (Sandoa, 1958), APK/ATS.

Labrique. *Exposé de la guerre Lunda-Tshokwe* (5 March 1922), AMC.

Mbundj, Oscar. *Note sur la succession du Grand Chef Mwanta Yamvo* (Kapanga 1 July 1963), APK.

―――. *Oraison funèbre de Mwanta Yamvo Mbako Ditende Yawa Naweji III* (Kapanga, 2 June 1963), APK.

Montenez, Prosper. *Historique de l'organisation administrative de l'ex-District de la Lulua* (20 January 1933), AMC.

Mwana-Kasongo, Francois-Xavier. *Rapport sur l'investiture du Grand Chef Mwanta Yamvo Mushidi* (Kapanga, 18 November 1963), APK.

Rapport de la mission géographique du Lac Dilolo (Mission Willemoes) (16 March 1907), AMC.

Scarambone, Lt. *Rapport politique sur la situation générale du poste de Katola* (Katola, 21 September 1904), AMC.

Simonatti. *Rapport sur la reconnaissance effectuée par M. l'Adjoint Supé-*

rieur Simonatti dans le Territoire de Kayoyo, District de la Lulua (May-July 1916, 2 parts), AMC.

Van de Byvang, M. *Administration indigène chez les Tutshiokwe* (n.d.; 1926?), AMC.

———. *Conclusions à l'étude des Aluunda et des Tutshiokwe* (n.d.; 1926?), AMC.

———. *Organisation sociale chez les Aluunda* (n.d.; 1926?), AMC.

Van Not, L. *Notice concernant les Lunda et les droits du Mwata-Yamvo sur des Bakete habitant le District du Kasai* (14 June 1919), AMC/ATK.

Verdick, Edgard. *Notes du Commandant Verdick (1er voyage au Congo)* (1890), AMC.

———. *Rapport sur la fondation des postes du Sud* (17 August 1903), AMC.

Waldecker, B. *Note au sujet du traitement du Mwant-Yav des a-Ruund* (Elisabethville, 7 December 1962), APK.

Correspondence files: APK, ATK.

Nonofficial material

Published

Anstey, Roger. *King Leopold's Legacy* (London, Oxford University Press, 1966).

"A propos des problèmes du milieu rural congolais." *Bulletin du CEPSI,* March 1957, no. 36.

Apter, David E. *The Political Kingdom in Uganda. A Study of Bureaucratic Nationalism* (Princeton, N. J., Princeton University Press, 1961).

Arnot, Frederick Stanley. *Bihé and Garenganze* (London, Hawkins, 1893).

A.T.C.A.R. "Programme," *L'Essor du Congo,* 16 September, 1959.

Austen, Ralph A. *Northwest Tanzania under German and British Rule* (New Haven, Conn., Yale University Press, 1968).

Bateman, C. S. Latrobe. *The First Ascent of the Kasai* (London, G. Phillip and Son, 1889).

Batista, Pedro João. "Explorações dos Portugueses no interior d'Africa meridional. . . . Documentos relatives a viagem de Angola para Rios de Sena," *Annães Maritimos e Coloniaes,* III:5-7 and 9-11 (1843).

Baumann, Herrmann. *Lunda: Bei Bauern und Jägern in Inner Angola* (Berlin, Würfel Verlag, 1935).

Biebuyck, Daniel. "Fondements de l'organisation politique des Lunda du Mwaantayaav en Territoire de Kapanga," *Zaïre,* XI:8 (1957), 787-817.

Birmingham, David. "The Date and Significance of the Imbangala Invasion of Angola," *Journal of African History,* VI:2 (1965), 143-152.

Birmingham, David. *Trade and Conflict in Angola: The Mbundu and their neighbors under the influence of the Portuguese, 1483-1790* (Oxford, Clarendon Press, 1966).

Boone, Olga. *Carte ethnique du Congo: Quart Sud-Est.* Tervuren: MRAC, 1961

Brau, Camille. "Le droit coutumier Lunda," *Bulletin des Juridictions Indigènes et du Droit Coutumier Congolais*, vol. X (1942) no. 8, 155-176; no. 9, 179-203; no. 10, 205-229; no. 11, 231-252; no. 12, 255-267.

Brausch, Georges. *Belgian Administration in the Congo* (London, Oxford University Press, 1961).

Büchner, Otto. "Die Büchner'sche Expedition." *Mitteilungen der Afrikanischen Gesellschaft Deutschlands*, I:4-5 (1879); II:1, 3 and 4 (1880); III:1 (1891).

Büchner, Max. "Das Reich des Muatiamvo und seine Nachbarländer," *Deutsche Geographische Blätter*, VI (1883), 56-67.

Bustin, Edouard. "Government Policy Toward African Cult Movements: A Case Study from Katanga," *African Dimensions: Essays in Honor of William O. Brown*, ed. Mark Karp (New York, Africana Publishing Corp., 1975).

Capello, H. de Brito and Roberto Ivens. *De Angola à contra-costa* (Lisbon, Imprensa Nacional, 1886).

_____. *De Benguella às terras de Iàccà* (Lisbon. Imprensa Nacional, 1881).

Carvalho, Henrique A. Dias de. *Descripção da viagem à Mussumba do Muatiânvua* (Lisbon, Imprensa Nacional, 1890-1894), 4 vols.

_____. *Ethnographia e Historia tradicional dos povos da Lunda* (Lisbon. Imprensa Nacional, 1890).

_____. *O Jagado de Cassange* (Lisbon, 1898).

_____. *O Lubuco* (Lisbon, 1889).

_____. *A Lunda; ou, os estados do Muatiânvua: Dominos da soberania de Portugal* (Lisbon, Adolpho Modesto Impressores, 1890).

Centre de Recherche et d'Information Socio-Politiques. *Congo 1959;* id. 1960 through 1967 (Bruxelles, CRISP, 1960-1970).

Chinyanta Nankula, Mwata Kazembe XIV. *Historical traditions of the Eastern Lunda,* trans. Ian Cunnison (Lusaka, Rhodes-Livingstone Institute, 1961).

_____. *My Ancestors and my People* (London, 1951).

Colvin, Ian. *The Rise and Fall of Moïse Tshombe* (London, Frewin, 1968).

Le Congo Belge (Bruxelles, INFORCONGO, 1958), 2 vols.

Cornet, René J. *Le Katanga avant les Belges et l'expédition Bia-Francqui-Cornet* (Bruxelles, L. Cuypers, 1946).

Coussement, Grégoire. "La secte du Punga et du Mama Okango," *Bull. des Juridictions Indigènes et du Droit Coutumier Congolais,* May 1935, pp. 64-67.

Crine, Fernand. "Aspects politico-sociaux du système de tenure des terres des Luunda septentrionaux," in *African Agrarian Systems,* ed. D. Biebuyck

(London; Oxford University Press for the International African Institute, 1963).

Crine, Mavar, B. "Histoire traditionnelle du Shaba," *Cultures au Zaïre et en Afrique,* no. 1 (1973), 5-108.

Crowley, Aiden. "Politics and Tribalism in the Katanga," *Western Political Quarterly,* XVI:1 (1963), 68-78.

Cunnison, Ian. "History and Genealogies in a Conquest State," *American Anthropologist,* LIX (1957), 20-31.

————. *History on the Luapula* (Capetown, Oxford University Press, 1951).

————. "Kazembe and the Portuguese, 1798-1832," *Journal of African History,* II:1 (1961), 61-76.

————. *The Luapula Peoples of Northern Rhodesia: Custom and History in Tribal Politics* (Manchester, Manchester University Press, 1959).

————. "Perpetual Kinship: A Political Institution of the Luapula Peoples," *Rhodes-Livingstone Journal,* 20 (1956), 28-48.

————. "A Watchtower Assembly in Central Africa," *International Review of Missions,* XL, no. 160 (1951).

de Hemptinne, J. "La politique indigène du gouvernement belge," *Congo,* vol. 9, II:3 (October 1928), 359-374.

————. "Précisions sur le problème de la politique indigène," *Congo,* vol. 10, II:2 (Juillet 1929), 187-207.

de Jonghe, Edouard. "Formation récente de sociétés secrètes au Congo Belge," *Africa,* IX:1 (1936), 56-63.

Delcommune, Alexandre. *L'Avenir du Congo Belge menacé* (Bruxelles, J. Lebègue, 1921).

Delvaux, Henri. *L'occupation du Katanga, 1891-1900* (Elisabethville, Imbelco, 1950).

Denis, Jacques. "Elisabethville: Matériaux pour une étude de la population africaine," *Bulletin du CEPSI,* no. 34 (1956), 137-195.

————. "Jadotville: Matériaux pour une étude de la population africaine," *Bulletin du CEPSI,* no. 35 (1956), 25-60.

Denolf, Prosper. *Aan de Rand van de Dibese* (Bruxelles, IRCB, 1954).

de Sousberghe, L. "Les Pende: Aspects des structures sociales et politiques," in *Miscellanea Ethnographica* (Tervuren, MRAC, 1963).

de Vos, Stan (Msgr.). "La politique indigène et les missions catholiques," *Congo,* vol. 4, II:5 (December 1923), 635-650.

Dos Santos, Eduardo. *A Questão da Lunda (1885-1894)* (Lisbon, Agencia Geral do Ultramar, 1966).

Douglas, Mary T. "Matriliny and Pawnship in Central Africa," *Africa* XXXIV: 4 (1964), 301-313.

Dumont, Georges Henri. *La Table Ronde Belgo-Congolaise* (Bruxelles and Paris, Editions Universitaires, 1961).

Duysters, Léon. "Histoire des Aluunda," *Problèmes d'Afrique Centrale,* 1958, no. 40, 75-98.

Fetter, Bruce S. "The Luluabourg Revolt at Elisabethville," *African Historical Studies,* II:2 (1969), 269-276.

La Force Publique de sa naissance à 1914. (Bruxelles, IRCB, 1952).

Franck, Louis. "La politique indigène, le service territorial et les chefferies," *Congo,* vol. 2, I:2 (1921), 189-201.

P. G. "L'administration des indigènes," *L'Essor du Congo,* 14 June 1945.

Gamitto, Antonio Candido Pedroso. *O Muata Cazembe e os povos Maraves, Chevas, Muizas, Muembas, Lundas e outros da Africa Austral* (Lisbon, Agencia Geral das Colónias, 1937).

Gardet, Georges. *Histoire du Katanga* (Bruxelles, Eds. Répertoire de l'Industrie Universelle, 1913), 2eme ed.

Gérard-Libois, Jules. *Katanga Secession* (Madison, Wis., University of Wisconsin Press, 1966).

Graça, Joaquim Rodrigues. "Descripção da viagem feita de Loanda com destino as cabeceiras do Rio Sena," *Annães do Conselho Ultramarino* (Parte não oficial), ser. I (Feb. 1954-Dec. 1958).

_____. "Expedição ao Muatiâvnua: Diario de Joaquim Rodrigues Graça," *Boletim da Sociedade de Geographia de Lisboa,* IX:8-9 (1890), 365-468.

Grévisse, F. "Notes ethnographiques relatives à quelques populations autochtones du Haut-Katanga industriel," *Bulletin du CEPSI,* 32-41 (March 1956 to June 1958).

_____. "Les traditions historiques des Basanga et de leurs voisins," *Bulletin du CEPSI,* 2 (1946-1947).

Guébels, Leon. *Relation complète des travaux de la Commission Permanente pour la Protection des Indigènes, 1911-1951* (Gembloux, Belgium, Duculot, n.d. (1952?).

Heussler, Robert. *The British in Northern Nigeria* (London, Oxford University Press, 1968).

Janssens, E. and A. Cateaux. *Les Belges au Congo* (Anvers, J. Van Hille-de Backer, 1908-1912), 4 vols.

Jentgen, Pierre. *Les frontières du Congo Belge* (Bruxelles, IRCB, 1952).

Kottak, Conrad P. "Ecological Variables in the Origin and Evolution of African States: The Buganda Example," *Comparative Studies in Society and History,* vol. 14 (1972), 351-380.

Labrecque, E. "Histoire des Mwata Kazembe, chefs Lunda du Luapula, 1700-1945," *Lovania,* vol. 16 (1949), 9-33; vol. 17 (1950), 21-48; vol. 18 (1951), 18-67.

_____. "La tribu des Babemba," *Anthropos,* vol XXVIII (1933).

Lacerda, José de. *Exame das viagens do Dr. Livingstone* (Lisbon, 1867).

Lerbak, Anna E. and Daniel Munuung, eds. *Ngand Yetu: Uruund wa Mwant*

Yavu (Elisabethville, Eglise Méthodiste, 1963).

Liebenow, J. Gus. *Colonial Rule and Political Development in Tanzania: The Case of the Makonde* (Evanston, Ill., Northwestern University Press, 1971).

Liebrechts, Charles. *Léopold II, fondateur d'empire* (Bruxelles, Office de Publicité, 1932).

Livingstone, David. *Missionary Travels and Researches in South Africa* (London, John Murray, 1857).

Lombard, Jacques. *Structures de type "féodal" en Afrique noire: Etude des dynamismes internes et des relation sociales chez les Bariba du Dahomey* (Paris and The Hague, Mouton, 1965).

Low, D. Anthony and R. Cranford Pratt. *Buganda and British Overrule, 1900-1955* (London, Oxford University Press, 1960).

Lucas, Stephen A. "The Outsider and the Origin of the State in Katanga," Mimeographed paper, African Studies Association Annual Conference, 1971.

Malengreau, Guy. *Vers un paysannat indigène* (Bruxelles, Van Campenhout, 1949).

Maquet, Jacques J. *Afrique: Les Civilisations noires* (Paris, Horizons de France, 1960).

———. "Une hypothèse pour l'étude des féodalites africaines," *Cahiers d'Etudes Africaines,* vol. II, no. 6 (1962), 292-314.

Marcum, John. *The Angola Revolution,* vol. 1 (Cambridge, Mass., M.I.T. Press, 1969).

Margarido, Alfredo. "La capitale de l'Empire Lunda: un urbanisme politique," *Annales,* XXV:4 (1970), 857-861.

———. "Processus de domination formant un empire: le cas des Lunda," *Présence Africaine,* XXVII, 55 (1965), 100-117.

McCulloch, Merran. *The Southern Lunda and Related Peoples* (London, International African Institute, 1951).

Merlier, Michel. *Le Congo, de la colonisation belge à l'indépendance* (Paris, Maspero, 1962).

Michaux, Oscar, *Carnets de Campagne (1889-1897)* (Namur, Dupagne-Counet, 1913).

Miller, Joseph Calder. *Cokwe Expansion, 1850-1900* (Madison, Wis., University of Wisconsin, African Studies Program, 1969).

Miracle, Marvin. *Agriculture in the Congo basin* (Madison, Wis., University of Wisconsin Press, 1967).

Moloney, James A. *With Captain Stairs to Katanga* (London, S. Low, Marston and Co., 1893).

Montenez, Prosper. "Notes sur l'identité coutumière des indigènes d'origine

Lunda," *Bulletin des Juridictions Indigènes et du Droit Coutumier Congolais,* IV:11 (1936), 269-277.

Mulambu-Mvuluya, Faustin. "Contribution à l'étude de la révolte des Bapende," *Cahiers du CEDAF,* 1971, no. 1.

———. "Cultures obligatoires et colonisation dans l'ex-Congo Belge," *Cahiers du CEDAF,* 1974, no. 6-7.

"Mutineries au Congo Belge," *Bulletin Militaire,* March 1947.

Mwepu-Kyabutha, Gaspard. "Quelques aspects des conséquences sociales de l'industrialisation du Katanga," *Civilisations,* XVII:1-2 (1967), 53-68.

Nicolai, Henri. *Le Kwilu: Etude géographique d'une région congolaise* (Bruxelles, CEMUBAC, 1963).

Nsaang ja Aruund. n.p. (Musumba), n.d. (1970?).

Plancquaert, M. *Les Jaga et les Bayaka du Kwango* (Bruxelles, IRCB, 1932).

Pogge, Paul. *Im Reiche des Muata Jamwo* (Berlin, Verlag Dietrich Reimer, 1880).

Redinha, José. *Etnosociologia do Nordeste de Angola* (Lisbon, Agencia Geral do Ultramar, 1958).

———. "Historia e cultura dos Quiocos da Lunda," *Mensario Administrativo* (Luanda), 18:7-9 (1964) 33-46.

Richards, Audrey I. "The Bemba of Northeastern Rhodesia," in *Seven Tribes of British Central Africa,* eds. E. Colson and M. Gluckman (Manchester, Manchester University Press, 1951).

Roland, Hadelin. "Résumé de l'histoire ancienne du Katanga," *Bulletin du CEPSI,* 61 (June 1963), 3-41.

Rubbens, Antoine, *et al. Dettes de guerre* (Elisabethville, Editions de L'Essor du Congo, 1945).

Ryckmans, Pierre. *Etapes et Jalons* (Bruxelles, F. Larcier, 1946).

Schütt, Otto H. *Reisen im Sudwestlichen Becken des Congo* (Berlin, Dietrich Reimer, 1881).

Shils, M. "Enquête politico-foncière, Province du Katanga, District du Lualaba, Territoire de Dilolo," *Bulletin des Juridictions Indigènes et du Droit Coutumier Congolais,* XXV:4 (1957), 99-109.

Sohier, Jean. "Institutes coutumières katangaises," *Problèmes Sociaux Congolais,* no. 62 (September 1963), 3-83.

Southall, Aidan. *Alur Society: A Study in Processes and Types of Domination* (Cambridge, Heffer, 1956).

———. "Belgian and British Administration in Alurland," *Zaïre,* VIII:5 (1954), 467-486.

Springer, John McKendree. *Christian Conquests in the Congo* (New York, Methodist Book Concern, 1927).

Stevenson, Robert F. *Population and Political Systems in Tropical Africa*

(New York, Columbia University Press, 1968).

Struyf, Ivo. "Kahemba: Envahisseurs badjok et conquérants balunda," *Zaïre*, II:4 (1948), 351-390.

Tippu Tib (Hamed Bin Muhammed el Murjebi). "Maisha ya Hamed bin Muhammed El Murjebi, yaani Tippu Tib," transl. by W. H. Whiteley (Kampala, East African Literature Bureau, 1970).

Toussaint, R. "Notes sur les redevances en matière judiciaire prévues par la coutume en Territoire de Kapanga," *Bulletin des Juridictions Indigènes et du Droit Coutumier Congolais*, XXVI:11 (1958), 321-323.

Turner, Victor W. "A Lunda Love Story and its Consequences," *Rhodes-Livingstone Journal*, XIX (1955), 1-26.

————. *Schism and Continuity in an African Society* (Manchester, Manchester University Press, 1957).

Tylden, G. "The Gun Trade in Central and Southern Africa," *North. Rhodesia Journal*, II:1 (1953), 43-48.

Union Minière du Haut Katanga (Bruxelles, L. Cuypers, 1956).

Van Bulck, Gaston. *Les recherches linguistiques au Congo Belge* (Brussels, IRCB, 1948).

Van den Byvang, M. "Notice Historique sur les Balunda," *Congo*, vol. 18 (1937) I:4, 426-438; I:5, 548-562; II:2, 193-208.

Van der Kerken, Georges. *La politique coloniale belge* (Anvers, Editions Zaïre, 1943).

Van Roy, H. "L'origine des Balunda du Kwango," *Aequatoria*, XXIV:4 (1961), 136-141.

Vansina, Jan. "La fondation du royaume de Kasanje," *Aequatoria*, XXV:2 (1963), 45-62.

————. *Introduction à l'ethnographie du Congo* (Kinshasa, Editions Universitaires du Congo, 1966).

————. *Kingdoms of the Savanna* (Madison, Wis., University of Wisconsin Press, 1966).

————. "Long-distance trade routes in Central Africa," *Journal of African History*, III:3 (1962), 375-390.

————. "More on the Invasions of Kongo and Angola by the Jaga and the Lunda," *Journal of African History*, VII:3 (1966), 421-429.

————. *Oral Tradition* (London, Routledge and Kegan Paul, 1965).

————, Raymond Mauny, and L. V. Thomas, eds., *The Historian in Tropical Africa* (London, Oxford University Press 1964).

Van Zandijcke, A. "Note historique sur les origines de Luluabourg (Malandi)," *Zaïre*, VI:3 (1952), 227-248.

————. "La révolte de Luluabourg," *Zaïre*, IV:9 (1950).

Van Zuylen, P. *L'Echiquier congolais ou le secret du Roi* (Bruxelles, Dessart, 1959).

Vellut, J. L. "Notes sur le Lunda et la frontière luso-africaine, 1700-1900," *Etudes d'Histoire Africaine,* vol. III (1972), 61-166.

Verbeken, Auguste. *Contribution à la géographie historique du Katanga et des régions voisines* (Bruxelles, IRCB, 1954).

_____. *Msiri, roi du Garenganze: l'homme rouge du Katanga* (Bruxelles, L. Cuypers, 1956).

_____, and M. Walraet. *La première traversée du Katanga en 1806* (Bruxelles, IRCB, 1953).

Verdick, Edgard. *Les Premiers jours au Katanga, 1890-1903* (Bruxelles, Editions du Comité Spécial du Katanga, 1952).

Verhulpen, Edmond. *Baluba et balubaïsés du Katanga* (Anvers, Eds. de l'Avenir Belge, 1936).

Vermeulen, Victor. *Déficiences et dangers de notre politique indigène* (Bruxelles, Impr. Veuve Monnom, 1952).

_____. "Les 'villages modèles' de la Lulua (1928-1934): Une expérience dans le domaine du bien-être indigène," *Revue Coloniale Belge,* no. 79 (15 January 1949).

Wauters, A. J. "Chez le Muata Yamvo," *Mouvement Géographique,* no. 40 (1897), 470.

_____. "Les prétentions portugaises sur le Mouata Yamvo," *Mouvement Géographique,* no. 22 (1890), 91.

Weiss, Herbert. *Political Protest in the Congo* (Princeton, N. J., Princeton University Press, 1968).

White, C.M.N. "The Balovale People and Their Historical Background," *Rhodes-Livingstone Journal,* no. 8 (1949), 26-41.

_____. "The Ethno-History of the Upper Zambezi," *African Studies,* XXI:1 (1962), 10-27.

_____. *An Outline of Luvale Social and Political Organization* (Manchester, Manchester University Press, 1960).

Willame, Jean-Claude. *Patrimonialism and Political Change in the Congo* (Stanford, Cal., Stanford University Press, 1972).

_____. *Les provinces du Congo,* vol. 2, *La Province du Sud-Kasai* (Léopoldville, Université Lovanium, 1964).

Willequet, J. *Le Congo Belge et la Weltpolitik* (Bruxelles, Presses Universitaires de Bruxelles, 1962).

Wissman, Hermann von. *My Second Journey Through Equatorial Africa from the Congo to the Zambezi in the Years 1886 and 1887* (London, Chatto and Windus, 1891).

_____. *Unter Deutscher Flagge: Quer durch Afrika von West nach Ost, 1880-1883* (Berlin, Walther und Apolant, 1889).

Young, M. Crawford. *Politics in the Congo* (Princeton, N. J., Princeton University Press, 1963).

Zolberg, Aristide. *Creating Political Order: The Party States of West Africa* (Chicago, Rand McNally, 1966).

Unpublished

Bastin, Marie-Louise. "Tshibinda Ilunga, héros civilisateur," unpubl. thesis, Université Libre de Bruxelles, 1966.

Hoover, J. Jeffrey. "Over, Under, Around but not Through: A Comparative Study of the Lunda Under Belgian and British Rule," seminar paper, Yale University, Dept. of History, 1970.

Lucas, Stephen A. "Baluba et Aruund: Etude comparative des structures sociopolitiques," unpubl. diss., Paris, Ecole Pratique des Hautes Etudes, 1968.

Muhunga, Ambroise. "La grande nation des Tshokwe et sa dynastie," (undated MS), MRAC.

Muyere-Oyong, N. "Promotion des collectivités locales en République du Zaïre. Etude des dynamismes des autorités traditionnelles appliquée au cas des chefferies du Mwaant Yaav et des Bayeke," unpubl. diss., Université Libre de Bruxelles, 1973.

N'dua, Edouard. "L'installation des Tutshokwe dans l'Empire Lunda, 1850-1903," unpubl. thesis, Kinshasa, Université Lovanium, 1971.

Strythagen, R. "Note pour les membres du Comité Local de l'Office Central du Travail au Katanga, no. 10" (Elisabethville, 5 January 1931), APK.

Private Correspondence.

Index